HOW OUR BRAIN BECAME HUMAN

Genes, Environment, Microbiome, Social Life and their Interactions

Yanko A. Yankov, MD, PhD.

Order this book online at www.trafford.com
or email orders@trafford.com

Most Trafford titles are also available at major online book retailers.

Print information available on the last page.

ISBN: 978-1-6987-1310-6 (sc)
ISBN: 978-1-6987-1311-3 (hc)
ISBN: 978-1-6987-1309-0 (e)

Library of Congress Control Number: 2022918520

Trafford rev. 02/13/2023

North America & international
toll-free: 844-688-6899 (USA & Canada)
fax: 812 355 4082

CONTENTS

Dedication

To my parents, Dr. Athanas and Dr. Petia Yankov, who gave my sister, Sophie, and me daily lessons of courage, curiosity, persistence, and love with their exemplary lives and dental work

ACKNOWLEDGMENTS AND SELF-INTRODUCTION

I am a lucky and inquisitive man. Born and raised in socialistic Bulgaria, somehow, I saw only the beauty around me. My continuous medical curiosity and the fiftyplus years of studying and treating patients with neurological, genetic, and sleepdisorders have motivated me to write this book. The picture below of some of our professors in the Medical Academy, Sofia, was taken in 1984, ten years after I graduated from that academy.

I remember all my professors by name, and the lessons they taught me have stayed engraved in my memory. My first mentor, Dr. Christo Koltchev (below), introduced me to electroencephalography (EEG) and sleep medicine. Later, we develop a long and wonderful friendship.

In 1981, I defended my PhD thesis, "Carrier Detection in Duchenne and Becker Muscular Dystrophies," under the supervision of Professor Elena Tzvetanova. Shortly after that, I was offered the position of assistant professor in my alma mater, the Neurology Clinic, Sofia. In 1986, I met a cheerful and intelligent young woman, Ina, and we married. Our children, Svetla and Petia, grew up to become responsible scientists and entrepreneurs.

In 1992, Dr. Alexander Todorov, from Tuscaloosa, Alabama, invited me to join his neurology clinic. There, I met my wonderful second wife, Jo Ellen. She continues to introduce me to American culture and customs.

After taking all my medical exams, I was lucky again to be accepted at age forty-five for my internship and residency at the University of Texas Health Science Center at San Antonio (UTHSCSA) under the tutorship of Professor Dr. David Sherman.

Thanks to him and all the staff of the Neurology Department, I accumulated more clinical knowledge and experience. Later, this knowledge was enriched during my EEG and Sleep Medicine Fellowship at the Neurophysiology Department of UTHSCSA under the supervision of Dr. Szabo, Dr. Mayes, Dr. Thomas, Dr. Cavazos, Dr. Ingmundson, and Dr. Jackson.

In December 2001, by invitation of Dr. Peter Tarbox and Dr. Robert and Dr. Lynell Lowry, I joined their private practice, Neurology and Neurophysiology Clinic, in San Antonio. I also took the position of medical director of the Alamo Sleep Center, where Dr. Paul Ingmundson, the clinical director of this center, became my first teacher in sleep medicine. I am grateful also to John and Margaret Rackler, who were the business managers of the center. My other teacher of sleep medicine is Dr. James Andry, who became, along with Dr. Ingmundson, my precious friends too. During the next ten years, I saw, diagnosed, and treated many patients with different neurological and sleep disorders. At the same time, I passed the national certification exams by the Boards of Neurology, Clinical Neurophysiology, and Sleep Medicine. In 2012, I started my own neurology and sleep medicine practice in Bulverde, close to San Antonio.

Throughout my entire medical career, I continuously shared clinical and scientific experiences with my colleagues, other medical providers, and patients. For the last twenty years, our small medical and scientific community has continued to meet every Sunday for coffee and discussions. All my gratitude to all of you, my friends, for the exchange of ideas and for keeping me updated in the novelties in medicine, science, and life. You are my extended family: Vroni, Dimitry, Nina, Carmen, Josep, Malgorjata, Ed, Lawrence, Ibrahim, Eddie, Sam, Kathleen, Richard, Wiley, Sameh, Aaron, Diana, Swann, Suzan, Marina, Scott, Julie, Dallas, Michael, Moffatt, Zach, Megan, and many others. You all have richly contributed to my life and this book.

Thank you, Florence Weinberg from San Antonio, for being my adviser and editor and for taking your "job" with scientific passion.

Thank you, my young cover page illustrator, Natalie, a talented aspiring medical student, presently a scribe for our primary MD, Dr. Carmen Cawley.

I profoundly thank my wife and family, colleagues, teachers, friends, and patients, who helped me continuously study this miracle called the *brain*. I thank especially my ex-wife, Ina Yankova, for her deep interest in my writings and her valuable remarks.

INTRODUCTION

The unexamined Life is
not worth living.
—Socrates

An immense body of literature has been written on human brain growth and evolution. The subject is very complex and probably beyond the reach of a single person. I am also aware of the multitude of opinions regarding how we came to exist and why we are here on earth. I would like this book to allow us to freely exchange our scientific knowledge and different opinions. I will try to systematically and clearly present the multidisciplinary scientific data currently known about human brain growth and evolution. I will do my best to review the literature that discusses this multifaceted, fundamental, and permanent question many of us ask. As a clinical neurologist, neurophysiologist, and sleep specialist, I do not have my own experimental data but will present some data from my fifty years of clinical experience. The review of the literature will include more than a hundred books and articles related to the problem. Since the information on human brain growth and evolution is so vast and complex, I will divide the book into seven main chapters:

The first chapter will give an overview of the known scientific knowledge about the evolution of the ecosystems of our planet, from its formation, estimated at 4.54 billion years ago, to the present day. Naturally, the focus will be on the evolution of primates and their brains. Science teaches that our common ancient ancestors, primates and monkeys, branched off from a big group of mammals approximately twenty-five million years ago. Note: Monkeys have a tail; primates do not. The interesting fact is that the name "chimpanzee" first appeared in 1920 from one bonobo and one chimpanzee named Chim and Panzee. These two primates were purchased by a then famous primatologist, Robert Yerkes. Chim and Panzee lived happily with the primatologist, who derived the name "chimpanzee" from the combination of their given names.

The second chapter will be a presentation of the known facts on gene mutations, gene duplications, and other gene changes that are unique to our human brain when compared to the chimpanzee that took place over the last six to seven million years. Current science holds that our common ancestors lived at that time alongside the chimpanzee. Opinions on this are very diverse. As you may know, there are on average four thousand medical science publications every day printed in medical journals and numerous books. It is not possible to cover the vast literature on genetic and other factors influencing human brain growth and evolution. I apologize in advance to those authors who have written on this matter, whose findings I could not include in the literature review. This chapter will review the twelve best-known specific gene mutations involved in human brain growth as compared to our closest genetic primate—the chimpanzee. I will mention only eighteen gene mutations probably also involved in this complex process. Interestingly, these specific human gene mutations are in general beneficial for growing more neurons and more complex glia, but at the same time, some of these genes seem easily breakable. Thus, these mutations, beneficial for most humans, can eventually develop some human-specific diseases like schizophrenia and autism in some individuals. Until now, the community of brain evolution scientists was focused mainly on six genes and their mutations, the first ones discussed in this chapter.

The third chapter will review the role of our continuously changing environment on the growth and increased complexity of our brain. This part of science is called epigenetics. The term "epigenetic" comes from Greek and means "in addition to the genes." This area of knowledge is relatively new and known mainly to dedicated specialists. In this part, I will discuss the following:

- The role of DNA methylation on gene expression
- The unique adaptability of humans to environmental changes, which can be attributed in part by the increased number of our sweat glands, bipedalism, and other body changes allowing humans to increased and safer exposure to the sun
- The discovery of fire and cooking and their impact on the early human brain growth. The routine use

of fire by our ancestor hominids reportedly started around 1.5 million years ago. The use of fire for cooking plays an enormous role in diversifying and enriching early humans' food supply and, consequently, providing more energy for brain growth and development

- The role of physical activity, stress, and sleep on brain development

The fourth chapter will review the role of some newly discovered neuropeptides and transcription factors as well as the role of glia in human brain evolution. The two relatively newly discovered hormones, trophic factors, and neuromodulators are named PACAP and VIP. They have evolved for the last five hundred million years and are part of the glucagon and secretin hormone family. You are probably hearing about them for the first time. VIP is the abbreviation of vasoactive intestinal polypeptide, and PACAP stands for pituitary adenylate cyclase activated polypeptide. The role of these neuropeptides for the immune and nervous systems is an object of present intensive research; still, they are poorly popularized even in medical literature. There are no publications about their role in human brain growth and evolution.

In this chapter, I will also address the topic of astroglia cells. They are the "long-forgotten" brain cells constituting approximately 85 percent of our brain. I will specifically draw your attention to these "forgotten" brain cells, along with other glial cells such as microglia (the "brain macrophages") and oligodendroglia (myelin-producing cells) and their role in human brain growth and evolution. In addition, this chapter will outline the role of ApoE genes and the key important nutrients for human brain evolution and human longevity as well as their role in disease processes such as Alzheimer's and other types of dementia.

The fifth chapter is dedicated to the importance of food and our gut microbiome in human brain growth and evolution. In the last decades, the gut microbiome has been increasingly researched mainly by functional medicine, while the conventional medical school still lags behind. In San Antonio, we are trying to fill this gap. For the last twenty-five years, we have our "international group," initiated by our friend and retired psychiatrist Dr. Vroni Hetherly. Our weekly medical discussions focus often on gut microbiome helping us understand the body as a whole entity and how to keep our body and mind in balance. I will also present some new data on microbiome-brain interactions, a focus of interest of more and more scientists and curious people from all professions. The few publications known about the role of the microbiome in human brain growth and evolution will be quoted.

The sixth chapter will discuss the key role of different human social factors: living in bigger communities, an active social life, communication (especially language), writing, beliefs, religions, cultural traditions and their place in society, and human brain development. This chapter will explain some of the complex interactions between the social factors and the biological ones mentioned in the previous chapters. Biological and social factors, as appeared, are heavily intermixed and probably equally important for the continuous genetic and social coevolution of our brains.

The interaction between genes, epigenetic factors, trophic factors, microbiome, and social life for human brain evolution is the main theme that I will follow throughout this book.

In the last, seventh chapter of the book, I will try to give my perspective on how our big and complex brains can help us live better individually and in the community. The better understanding of our brains will help us to live better as individuals and together in harmony with our environment. With the aid of continuously advancing technology, like CRISP-R, our human collaboration can help fight diseases and prolong our life and the life of our planet. Our children, our grandchildren, and their children deserve all our efforts to do so. The young Greta Thunberg from Sweden is a phenomenon that we follow closely, and I will be happy to contribute with this book to her and other movements trying to preserve our planet. The main goal of this book is to deepen our knowledge about how biological life appeared and how our brain became so complex. We present only the scientific knowledge without disparaging any opinion or other knowledge. On the contrary, I will be interested to discuss different opinions raised by interested readers. I hope this book will also help to focus on our obligation to save our planet and continue the miraculous *gift of life* for generations to come.

CHAPTER I

Our Planet's Story

It is essential to understand our brains in some detail if we are to assess correctly our place in this vast and complex universe that we see all around us.
—Francis Crick

Before confronting the genetic and other factors involved in human brain growth and evolution, I will briefly give you some generally accepted scientific facts about the history of our planet and its ecosystems. The material for this chapter was collected over the years, and it is impossible to name the numerous sources of this acknowledged scientific information. I will discuss the earth's ecosystems primarily in relation to brain growth and evolution in mammalians, primates, and humans.

According to a general scientific estimate, there are presently more than 1.5 million different animal species on earth, and 98 percent of them, like insects and worms, are invertebrates. Here is a brief overview of the present scientific knowledge of the evolution of our earth's ecosystems and the environmental factors which are fundamental to biological life on our planet.

Science accepts that our solar system has existed for about 4.54 billion years. Around 3.5 billion years ago, the first unicellular creatures appeared in the oceans. They are now called prokaryotic because they have no nucleus (*prokaryotic* in Greek means "pre-cellular"). Approximately two billion years ago, one small prokaryotic bacteria-like cell was incorporated in another bigger prokaryotic cell. Instead of being "digested and eaten" by the big cell, the two cells started to cooperate. The small bacteria-like prokaryotic cell used the food supply from the larger cell and provided the larger cell with energy. Today, we call these small bacteria-like structures mitochondria and the process symbiosis, meaning "living together." Nowadays, nobody can imagine life without mitochondria and symbiosis.

Until six hundred million years ago, life existed exclusively in the oceans and was represented only by prokaryotic cells. Science believes that the first primitive multicellular animal life appeared in the oceans during that period. We accept that the so-called Cambrian period occurred approximately 550 million years ago. It is named "Cambrian" because the Paleozoic rocks from this period were first studied in 1836 in Wales. Wales was called Cambria at that time. During the Cambrian period, oceanic algae and ocean-floor plants started their "terrestrial invasion." Some marine animals went after the plants out of the ocean, and the first amphibian and terrestrial creatures appeared on the surface of earth.

At this period, the oxygen level started to rise and the levels of carbon dioxide to decrease. The increase of atmospheric oxygen allowed the plants and animals on the earth to grow bigger and bigger. Some Cambrian and post-Cambrian creatures, called dinosaurs, became huge animals and suppressed the existence of smaller species. A few of the smaller animals survived, mostly underground. The first mammals are estimated to have appeared around three hundred million years ago. Science named the ancestors of these animals "synapsids," a group of reptiles that gave rise to mammals. Additionally, scientific facts proved that about 252 million years ago, the "Great Dying" occurred, also known as the Permian period. It is believed that during this critical period for our planet, 90 percent of the marine species and 70 percent of terrestrial species perished. The reason for the "Great Dying" is still an enigma, but it cleared the way for the dinosaurs to populate the earth. It is accepted that more than seven hundred species of dinosaurs existed at that time.

For the next two hundred million years or so, these dinosaurs (from Greek, meaning "terrible lizards") dominated our planet's animal life. At that time, only few mammals existed, mostly nocturnal and living most of the time underground. It was estimated that around sixty-five million years ago, a very large meteor fell on our planet. A big crater with a radius of 1.3 miles was found in the Gulf of Mexico, the presumed site where scientists believe this gigantic meteor struck our Mother Earth. This cosmic impact created such a devastation and a long-lasting cloud of dust around our planet that all the big animals, including all terrestrial dinosaurs, perished along with approximately 50 percent of all other species.

Some smaller animals might have survived, especially underground, including some mammals and birds. As we know, birds are descendants of small dinosaurs. Thank you, big meteor! We would not be here without your massive impact on the earth.

After all big dinosaurs perished, the small mammals evolved to have grasping hands and feet. They also developed binocular vision. The age of separation of primates from all other mammals is calculated to be twenty-five million years ago. Nowadays, we have more than four hundred species of living primates, the majority of which feel at home in trees, but some species, including humans, now live on the ground. Chimps, gorillas, and the hominin group of primates, often called the African apes, descended from an ancestor that resembles the Dryopithecus. Its Asian contemporary, Siva niches, gave rise to orangutans. Gibbons diverged earlier. Great apes and humans are grouped into a single family, the hominids. Humans belong to the hominin tribe, from which only one species remains today. That is us, you and me.

It was discovered by mutation-genetic research that hominins diverged from chimps around six to seven million years ago. The review of the literature showed that the hominin fossil record currently consists of twenty-five known species. Only a few of them belong to our ancestors, while many others became extinct without giving rise to new species. The oldest preserved mammalian skeleton of a lemur-like animal is dated forty-seven million years old. Some scientists believe that we and all primates came from a kind of lemur-like animal.[1]

[1] Picture published by David Derbyshire in Popular Science Journal, May 2009.

How can we comprehend that hundreds of species of primates, extinct or still existing now, including us, can descend from such lemur-like animals? I will do my best to unfold this complex story. My editor pointed out to my attention that the hind leg bones and feet of this animal are larger and heavier than its arms and hands, which may have had opposable thumbs. This means that the animal was built to bear more weight on its hind legs or, in other words, occasionally to walk or hop upright with the help of its long tail for balance.

Have patience, please, dear reader. It will be perfectly fine if you disagree or have other ideas or opinions. I am open to consider any new ideas or interpretation of the scientific knowledge presented so far.

The definition of the word "evolution" is "the process by which different kinds of living organisms are thought to have developed and diversified from earlier forms during the history of the earth." According to Lieberman, evolution is a dynamic process which is still occurring today. The first major transformation of hominins was their ability to walk on two feet—bipedalism. Over time, they became energy-savers and long-distance runners. The second transformation in the story of the human body was adding foods other than fruits and tubers. At that time, between four million and one million years ago, Australopithecus lived in Africa. The skeleton of Lucy exemplified one of them. She reportedly lived 3.2 million years ago. Lucy's incomplete skeleton was found in Ethiopia in 1974.[2] The skeleton was named Lucy because

[2] Daniel E. Lieberman, The Story of the Human Body. Evolution, Health and Disease. Vintage Books, 2013.

when Don Johnson glanced at the ulna sticking out of the ground, he and his teammates (archeologists) were listening to the Beatles's song "Lucy in the Sky with Diamonds."[3]

In Lucy's era, hominins needed twelve years to reach adulthood. *Australopithecus*, such as Lucy, were prodigious chewers; they ate tubers, seeds, and plants. Africa's climate continued to change and became cooler and drier with the expansion of open savannahs. The name "savannah" came from the Native American Taino language *zabana,* meaning "treeless plain." Compared with Lucy's skull, our skull and consequently our brains are three times as big, our bodies have long legs and short arms, and our faces have no snout.

Homo habilis existed between 2.3 and 1.4 million years ago but still had smaller skulls/brains. The brain of *Homo rudolfensis* was slightly bigger. *Homo erectus* was the first ancestor we can characterize as significantly human. Plant food still accounted for 70 percent or more of its diet.

Homo erectus first evolved in Africa around 1.9 million years ago. Part of this species endured in Asia until two hundred thousand years ago. They weighed in the range of 88 to 150 pounds and were between 122 and 185 centimeters high, which is like modern human proportions. *Homo erectus* had longer legs and was an exceptional long-distance runner at a moderate speed and a good hunter. When humans run, they cool by sweating, while most quadrupeds cool by panting, which they

---

[3] In 2007, my wife and I had the unique experience of seeing the Lucy skeleton in the Houston Museum of Natural Science's exhibition, "Lucy's Legacy: The Hidden Treasures of Ethiopia." This was only the third Lucy public appearance in nearly thirty years. Lucy's skeleton is estimated to be around 40 percent complete.

cannot do while galloping, except for horse, donkey, and probably the zebra. They sweat excessively. It is possible that the horse was domesticated by *Homo erectus* because, since it also sweated and could run even faster, it could exhaust the prey animal quicker.

The children of Homo erectus became fully adult at ages twelve to thirteen. We humans take an average of eighteen years to mature. About six hundred thousand years ago, *Homo erectus* had evolved into *Homo heidelbergensis*. This discovery was made at a single site in northern Spain, Sima de los Huesos. *Homo heidelbergensis* had a bigger brain of 1,100–1,400 grams and measured 143 centimeters in height. His body weight was 143–176 pounds. Archaic *Homo* invented the spearpoint and was able to control fire. *Homo heidelbergensis* moved south around the Mediterranean during the cold periods where they reconnected with their African ancestors. Molecular and fossil data indicates that *Homo heidelbergensis* diverged into several partially separated lineages four hundred thousand to three hundred thousand years ago. The African lineage evolved into modern humans. Another lineage evolved into the Denisovans in Asia.

The Denisovan is so named because part of the jawbone and one finger bone were found in the cave of Denis in the Altai mountains in Siberia. Denis was a Russian monk from the nineteenth century who lived for several years in this cave on a religious retreat. The Denisovan cave is also known as Aju-Tasch.

Another lineage of the archaic *Homo*, and the most famous one, found in Central and Eastern Europe, evolved from the Neanderthals. The Neanderthal was named because the first partial skeletons were found in caves near the village of Neanderthal, close to the Danube River, nowadays in Germany. The Neanderthal species was not formally recognized until 1863. Initially, it was erroneously considered to be our common ancestor. After 1945, scientists started to correctly recognize the Neanderthals as our "close cousins" who managed to survive in Europe during harsh glacial conditions. Recently, data showed that Neanderthals and modern humans are indeed separate species, diverged genetically at least eight hundred to four hundred thousand years ago. The Neanderthals' brains on average were bigger than the brains of modern humans. Their brain volume was 1,500 milliliters as compared to an average of 1,350 milliliters of modern humans. Nobody yet has explained why.

Modern humans can trace their roots to a common ancestral population that lived in Africa about three hundred to two hundred thousand years ago. That subset of humans started to disperse from Africa around eighty to one hundred thousand years ago. Modern humans appeared first in the Middle East between about one fifty and eighty thousand years ago but disappeared for about thirty thousand years when Neanderthals moved back into the region during the height of a major European glaciation, perhaps displacing humans for a while. Modern humans reappeared in the Middle East around fifty thousand years ago and rapidly spread from there in all directions. According to the majority data, modern humans appeared in Europe, Asia, and New Guinea forty thousand years ago. Soon after that, archaic humans soon became extinct. The latest remains of a Neanderthal, dating to thirty thousand years ago, was found in a cave in Spain.

The ice age ended 11,700 years ago. The climate became more stable without rapid, extreme climate changes. The predictable, consistent weather was good for hunter-gatherers and essential for farmers. Farmers and hunter-gatherers coexisted, often interbred, traded, and thus exchanged both genes and cultures.

According to David Christian, bipedalism, or the ability of our ancestors to walk on two feet, occurred three to four million years ago. Primates' brains are unusually large relative to their bodies, especially the cortex. Humans have as many as sixteen billion brain cortical neurons or more than twice as many as chimps. Chimps have only six billion brain cortical neurons. These data will be discussed later in the book with the appropriate citations. We appear to be very adaptable to survive but also very dangerous for other species. Our success story was not without significant "hiccups." One of these happened around seventy thousand years ago when the *Homo sapiens* population dropped to approximately seventy thousand people (close to extinction). It was believed that this was due to massive volcanic eruptions in present-day Indonesia. These eruptions covered the biosphere with dense clouds, blocking photosynthesis for months and years. Many societies starved, and many lives were destroyed, but there were survival champions, including us.[4]

Jen Viegas described the research of André Sousa and a group of coauthors who published in *Science Journal* why human brains are unique. The researchers investigated sixteen brain regions with 247 tissue samples in six humans, five chimpanzees, five macaque monkeys, and

---

[4] *Origin Story. A Big History of Everything.* NY: Little, Brown Spark, 2018.

other species. They found that the most distinct brain region is the striatum. The striatum is involved in motor coordination, reward, and decision-making.[5]

According to information from the Smithsonian National Museum, the difference between individual human genomes today is miniscule—about 0.1 percent on average; more details and citations in the next chapter. This means that the genetic material between any two individuals can differ at most only by 0.1 percent; the study of the same aspect of the human and chimpanzee coding genome indicates a difference of 1.2 percent—the same as with bonobos. The coding DNA difference between humans and gorillas is estimated at 1.6 percent. A difference of 3.1 percent distinguishes us and the African apes from the Asian great ape, the orangutan. A comparison of the entire genome, however, indicates that segments of DNA have also been deleted, duplicated over and over, or inserted from one part of the genome into another. When these differences are counted, there is an additional 4 to 5 percent distinction between human and chimpanzee genomes. It is known that we did not evolve directly from any primates living today. All humans and great apes differ from Rhesus monkeys, for example, by about 7 percent in their coding DNA.

---

5 *Jen Viegas, Mind, pp. 11/23/2017.*

On average, 85 percent of humans have the positive Rhesus factor in their blood. We abbreviate the Rhesus factor now as the Rh factor. Probably, most people on our planet talking about their Rh factor do not know that Rh is an abbreviation for Rhesus factor, present in the blood of the Rhesus monkey. Most people really do not know that 85 percent of us share this Rh factor with the Rhesus monkey. Yes, 85 percent of us are Rh (+), the humble author of this book included. I am simply blood group A (+). It is unclear too how most of the Basque people in the Pyrenees feel since most of them have a blood type without Rhesus factor or they are Rh (-). I simply know that we all should be proud to carry the brains of the only hominin survival champion, the master of our planet: *Homo sapiens.*[6]

[6] Dear reader, from now on, boxed sections like this one indicate that the enclosure contains the author's own contributions, not those of other authors and researchers cited in footnotes.

CHAPTER II
Genetic Changes Shaping the Human Brain

DNA is the canvas on which evolution works. The genome defines us, but does not make us.
—Unknown

2.1 General review of the role of genetic changes on human brain growth and evolution

The term "gene" was coined primarily by Danish botanist, plant physiologist, and geneticist Wilhelm Johannsen in 1909. Genes are organized in chromosomes. Between genes, there is a lot of genetic material and old retroviral residues of undetermined function called presently dark matter. All plants, microorganisms, and animals are composed from the same nucleosides that we all study in school and remember as RNA and DNA. Only few of us dedicate their lives to study genes and dark matter further. This chapter will be my attempt to synthesize the literature on this complex scientific quest.

Presently, according to Dr. Adam Rutherford, we have four living genera of Hominidae (also called great apes, no offense please):

- Pan (chimps and bonobos)
- Pongo (orangutans)
- Gorilla (gorilla)
- Humans

Chimps, bonobos, orangutans, and gorillas have twenty-four pairs of chromosomes. They have one more chromosome than we do. In humans, chromosome #2 is very big and was found to combine two chromosomes from the previous three primate groups. Dr. Rutherford continued in his book the chromosomal and biological analysis of our species. Neanderthals and Denisovans, like *Homo sapiens*, have twenty-three pairs of chromosomes. Neanderthals, Denisovans, and other yet undiscovered phantom cousins were different from us, but they also became part of us. Omnivorism was crucial to our survival. We are called omnivores because we "eat everything," meaning plants, fruits, meat, fish, seafood, etc. Our large, complex brain that kept us alive and for better or worse allowed us to influence the fate of the planet like no species ever before. The primary evolutionary driver of our diversity is gene duplication. Genetic studies reveal that since splitting with chimps, the human lineage has acquired almost seven hundred duplicated genes that chimps don't have. Many of these gene duplications are involved in brain function.[7]

---

[7] Adam Rutherford. A Brief History of Everyone Who Ever Lived. The stories of Our Genes. London: W&N, 2017.

I will enumerate and clarify all the known specifically human gene duplications and other gene mutations in the second part of the chapter.

The human genome was decoded for the first time in 2000 in a few individuals. The race to decipher the human genome is discussed in several books and many other sources. It is now publicly known that the primary decoding of our genome, a total of three billion pairs of nucleotides, costed more than three billion dollars! This decoding took ten years of hard work and innovations, involving many researchers divided into two competitive teams. Luckily, in the year 2000, the two competitive teams could shake hands in the office of President Bill Clinton and under the video watch of Great Britain's Prime Minister Tony Blair. As a result, many of us witnessed one of the most significant scientific achievements ever performed: the first decoding of the alphabet of our genome, containing three billion pairs of nucleotides. Thanks to the efforts of many scientists nowadays, everybody can have his or her genome decoded for a thousand dollars or less. What a century we are lucky to live in!

Matthew Cobb reported that on average, each of our human genomes differs from each other by about one base per thousand. In the same book, cited below, Cobb reported that the entire Neanderthal genome was decoded by Swante Paaabo and colleagues in 2010. They found that 2 to 3 percent of the European human genome is composed of Neanderthal genes. Some of the characteristics we carry from Neanderthals are related to skin color and a more robust immune response. The authors reportedly found approximately ninety genetic differences between humans and Neanderthals. Cobbs also reported that in 2011, Paabo published the genes of the Denisovan man. Like Neanderthals, they also inbred with humans. The findings displayed that Denisovans branched off from Neanderthals approximately three hundred thousand years ago. Apparently, Denisovan genes helped Tibetans survive the high altitude by adapting to lower oxygen levels.[8]

Despite the evident progress of research, the precise genetic changes between the brains of humans and other primates are largely unknown. John Copia believes that this is one of the greatest challenges in modern biology. These differences include our unique morphological traits, cognitive skills, spoken language, and disease susceptibilities. Special genetic studies identified 2,649 non-coding human accelerated regions, also known as HARs, that are unique to humans.[9]

> We can presume that no book or compendium can cover all these human-specific differences that made our brain unique, but why not keep trying.

Intelligence and intellectual processes in humans are vastly superior to those achieved by all other species. According to Rosales et al., brain evolution requires the coexistence of two adaptation mechanisms. The first mechanism involves genetic changes at the species level and the second at the individual level and involves changes

8 Matthew Cobb.; Life's Greatest Secret: The Race to Crack the Genetic Code. NY: Basic Books, 2019.

9 John A. Copia et al., Philos. Trans R. Soc., London Bibl. Sc., 12, 19, 2013.

in the chromatin aggregates or epigenetic changes. The genetic mechanisms include the following:

- Genetic changes in the coding regions that lead to changes in the sequence and activity of existing proteins
- Duplications and deletions of previously existing genes
- Changes in genetic expression through changes in the regulatory sequences of different genes
- Synthesis of non-coding RNAs (it is believed that we have 18,400 non-coding RNAs in our genome)[10]

Authors of an article in the journal *Science* (2008) asked this intriguing question: Why are our childhood and adolescence so long, not comparable to any other primates or mammals? Our big brain uses a lot of energy, especially with synapse formation and eliminating the old synapses, also called synapse pruning. Synapse pruning means "cutting off" and eliminating the old, unnecessary synapses by microglial cells in the brain. Lately, it was discovered that microglia recognize these synapses because they are marked by a particular antigen called complement Cq1 produced by astroglia cells.

The authors of the same article in *Science* wrote that humans are known for their big brains. Synapse pruning occurs mainly in the period between birth and twelve or thirteen years. During that time, most of our energy is concentrated in the brain. After age twelve and thirteen, the energy switches from the brain to the body and bone growth. Obviously, there is not enough energy for both processes simultaneously in humans because our brain

[10] M. A. Rosales et al. Neurologia, volume, 33, 4, 254–265, 2018.

is too big and consumes 20 percent of the energy we consume. Lately, some authors suspect that this separation of energy supply is responsible for our long childhood and adolescence. In all other mammals, both processes happen simultaneously. For the last five million years, human brains have tripled in size, with most of the brain growth occurring in the last two million years. The first fossil skulls of *Homo erectus* 1.8 million years ago had a brain volume averaging six hundred milliliters. From there, the species embarked on a slow upward march, reaching a brain volume of a thousand milliliters around five hundred thousand years ago. Early *Homo sapiens*, around two hundred thousand years ago, had brains averaging twelve hundred milliliters or more. This brain volume is within the range of the human brain size today.[11]

According to Swante Paabo, the genome between two people differs every twelve hundred to thirteen hundred nucleotide pairs, as opposed to one of every one hundred nucleotide pairs between humans and Neanderthals. Most nucleotide differences are seen in the seven hundred million people living in Africa. The name "Neanderthal" originated from a specimen found in 1856 in a limestone quarry in the Neanderthal area in Germany. The Neanderthal "technology" of making tools did not change for more than three hundred thousand years when they lived alone in Europe and part of Asia. By contrast, when modern humans populated the world outside Africa for the last hundred thousand years, technological tools have progressed to the present level. According to Paabo, there are eighty-seven different genes between modern humans and Neanderthals. Three of these genes are involved in

[11] *Science,* September 5th, 2008.

the formation of spindles of neural stem cell division, allowing more neurons to be born in humans. Every newborn baby has one hundred to two hundred new nucleotide mutations, out of our three billion nucleotide pairs, as compared to his/her parents.[12]

David Reich described the first whole genome of ancient human DNA established with the help of Dr. Paabo. Dr. Reich emphasized in his book that it is essential to understand our genetic past and how our interconnected human family was formed to identify risk factors for diseases.[13]

Presently, the above topic and many others are tackled daily by Dr. David Reich, Dr. Christopher Welch, Dr. Michael Greenberg, and others in their Human Brain Evolution Institute in Boston. This institution is receiving the support of the Paul G. Allen Institute, with headquarters located in Seattle, under the directorship of Dr. Christof Koch. The San Antonio Mind and Science Society was created by the visionary Texan rancher and entrepreneur Thomas Slick Jr., who also founded the Southwest Research Institute in San Antonio. The Mind and Science Society invited Dr. Christof Koch and other leading scientists such as Dr. Giulio Tononi for a two-day Symposium on Consciousness in 2017. We had the chance to hear and communicate with one of the best brain scientists in the world. Thank you, Mind and Science Society, for continuing and perfecting the legacy of Mr. Slick Jr. to combine science and technology.

---

[12] Swante Paabo, PhD. Video lecture for the Journal Sceptic, 2019. "Origins of the Humans" and in his book, Neanderthal Man, 2019.

[13] David Reich. *Who We Are and How We Got Here: Ancient DNA and the New Science of the Human Past,* Oxford University, 2018.

Above: Dr. Koch signing his book about consciousness during the Symposium on Consciousness organized by the Mind and Science Society in San Antonio, 2017

2.2 Individual gene effects on human brain growth and evolution

The second part of this chapter will review gene by gene the known unique human genetic changes making our brain bigger and more complex compared with nonhuman primates. Most of this information is so recent that it could be a novelty even for medical practitioners. However, despite the complexity of the genetic factors and the medical terminology, I appeal to you not to be discouraged. I will also add some of my own clinical observations in support of this literature review. Anyway, dear reader, keep reading and use your common sense. In the end, everything will fall into its place.

The first six genes below and their alterations in humans are relatively well known by specialists working in the field of human brain growth and evolution. Most of the other gene alterations are not well known. Their precise role in human brain evolution is not discussed even in very specialized medical literature. I will focus only on genes with a suspected role on human brain growth and evolution. At the end of the chapter, I will simply enumerate the sixteen genes that I think have the potential to be considered in the same category. With this simple gene enumeration, I will refrain, dear reader, from exposing you to many medical and genetic details that may contribute in making our brain bigger, more complex, or simply, human. Instead, I will try to simplify the matter, leaving space to younger researchers in their future scientific quests.

The NOTCH2NL gene

NOTCH2NL is a gene located on the first chromosome. This complex gene controls the timing of the development of all cells in every species. The focus will be placed only on the role of NOTCH2NL in the development of the human brain. According to Suzuki et al., the human-specific NOTCH2NL genes expand cortical neurogenesis which help the hominid cerebral cortex to undergo rapid expansion and increase in complexity.[14]

According to Ian Fiddes and coauthors, the gene NOTCH2NL is fully functional in human brains, nonfunctional in the brains of big primates like chimpanzees and gorillas, and completely missing in the macaque monkey. They also observed that the NOTCH

---

[14] *Suzuki, I. K. et al: Cell*, 173 (6), 1370–1384, May 31, 2018.

pathway is essential for radial glia stem cell proliferation and is a determinant in the number of cortical neurons in the mammalian cortex. They and other authors believe that the NOTCH simple genes existed in animals until twenty million years ago. The gene was named NOTCH because this gene gives rise to the notch from which insects' wings arise. According to Fiddes and coauthors, the first mutation of the NOTCH gene occurred in some primates fourteen million years ago. If science had existed at that time, this gene would be called NOTCH1NL. Nowadays, scientists know that at that time the gene was not yet functional.

According to the same publication, approximately three million years ago, the second duplication of the gene NOTCH happened but only in hominids. Nowadays, this gene is called NOTCH2NL. This gene became far more active in hominids, producing more stem radial cells and allowing more neurons to form. In nonhuman primates, the NOTCH2NL gene exists, but it is found to be inactive. Some authors believe that NOTCH2NL makes the human brain bigger but also more vulnerable to diseases.[15] According to Fiddes et al., NOTCH2NL genes provide the breakpoints in the 1q21.1 distal deletion/duplication syndrome. Duplications of the gene at that point are associated with macrocephaly and autism, and deletions are linked to microcephaly and schizophrenia. Thus, the human-specific NOTCH2NL gene may have contributed to the rapid evolution of the larger human neocortex, but alterations in the same gene could cause neurological and psychiatric disorders.[16]

---

[15] ibid.

[16] Ian T. Fiddes et al., online, in Cell, 10. 1016, May 31, 2018.

The SRGAP2 gene

The SPGAP2 abbreviation stands for SLIT-ROBO Rho GTPase-activating protein 2. This gene is located on the short arm of the first chromosome. The function of the SRGAP2 gene is to form dendritic spines which connect neurons to each other. The human SRGAP2 is duplicated. Mogen Y. Denis and coauthors also reported that similar duplicate genes found in humans represent approximately 5 percent of the human genome. These authors observed that one partial duplication of the SRGAP2 gene happened in hominids about 2.4 million years ago, known today as the SRGAP2C gene. The SRGAP2C gene increased the dendritic spines that forged connections with neighboring neurons.[17]

Fortua, Sikela, and coauthors discovered that the SPGAP2C gene had a third partial duplication in hominids approximately one million years ago, now called SRGAP2D. This gene formed even more neuronal dendrites. Currently, we all humans carry, on our first chromosome, the SRGAP2D gene, and we transfer it from generation to generation. The SRGAP2D gene is found only in humans, Neanderthals, and Denisovans.[18]

The ARHGAP11B gene

The ARHGAP11B abbreviation stands for Rho GTPase-activating protein. ARHGAP11B is located on the fifteenth chromosome. It encodes a protein containing 267 amino acids and has a GTPase activity which is

[17] *Dennis, M. Y. Cell*, vol. 149, issue 4, 912–922, May 2012.

[18] *PLOS Biology*, 2, e207, 2004.

essential for the brain stem cells, helping them to create our brain. This protein amplifies and controls neural cell production. This uniquely human protein also increases the folding of our cortex. ARHGAP11B was created by duplication of ARHGAP11A at or after our divergence from the chimpanzee lineage and before our separation from Neanderthals and Denisovans.

Recently, a team from the Max Planck Institute, led by Wieland Huttner, discovered the exact mutation in the gene of ARHGAP11B, leading to the human variant of the gene. They noticed that there is one point of mutation at the end of the gene consistent with a stretch of fifty-five nucleotides, where the nucleic base cytosine is replaced with another nucleic base: guanine. This point mutation, which happened presumably five million years ago, is found only in humans, Neanderthals, and Denisovans, but not in other nonhuman primates. This nucleotide substitution allows multiple divisions of the basal stem cells before they start their migration to the cerebral cortex. The ARHGAP11B gene is also involved in neuronal migration, neuronal differentiation, and synapse development. This gene increases the primary pool of neural stem cells, leading to more neurons landing in our cortex. In this way, our cortex is bigger and more folded than the cortex of other primates. In a video lecture, Dr. Huttner reported their findings with Martha Florio and Elena Taverna that the ARHGAP11B gene triples the basal progenitor cells migrating to the cortex. This primary neuronal pool is also enhanced by many other genes subject to discussion in this chapter.

The process of migration of the neurons from the periventricular area to the cortex starts in the embryo and is completed in humans by one year of age or so. This primary neuronal pool around the ventricles provides the excitatory and the inhibitory neurons for the cortex. The excitatory neurons represent approximately 90 percent of all neurons landing in the cortex; they use glutamate as a mediator. Their migration route to the cortex is direct and has been well known to science for the last thirty years. The other 10 percent of the cortical neurons are inhibitory and use GABA as a mediator. The migration of the inhibitory GABA neurons is much more complex and unclear in humans. Their migrations were just recently detailed in mice. These neurons seem to take an "oblique" road of migration to finally meet with the excitatory neurons in the brain's cortical areas.

During embryogenesis, the astrocytes are also "climbing" to the cortex. The astrocyte cells provide support and nutrition to the neurons. Astrocyte glial cells have been largely underestimated until recently. They were called glial cells when discovered at the beginning of the twentieth century by Ramon y Cajal and studied in detail from the 1920s by Ortega and many others. They were named glial cells because the scientists at that time believed they serve as a glue for the neurons (*glia* originates from the Greek meaning "glue"). The name "astro" was given because they look like "stars" under the microscope. I strongly advise my curious readers to read the excellent book of Douglas Fields, *The Other Brain*,[19] if they want to better understand the brain.

[19] R. Douglas Fields. *The Other Brain: The Scientific and Medical Breakthroughs That Will Heal Our Brains and revolutionize Our Health.* Simon & Schuster, 2009.

This book helped me understand the role of astroglia, oligodendroglia, microglia, and their complex relationships with the neurons. My interest in the glial cells' role in health and disease has kept increasing in the last several years. You can read more on the role of the glia for human brain development and function in chapter 3.

The ASPM gene

ASPM stands for the abnormal spindle-like microcephaly associated gene. The ASPM gene is found in the short arm of the 1st chromosome. The ASPM gene provides instructions for making a protein that is involved in cell division. This protein is found in cells and tissues throughout the body; however, it appears to be particularly important in the nervous system. Scientists have found that the ASPM protein helps maintain the orderly division of early brain cells called neural progenitors. The neural progenitor cells give rise to mature nerve cells (neurons) that travel like snails to the cortex. ASPM helps determine the total number of neurons and the overall size of the brain.

According to J. J. Zhang, the ASPM gene underwent an episode of accelerated sequence evolution in hominids after the split from chimpanzees, and a new allele of the ASPM gene appeared in humans sometime around six thousand years ago. Nowadays, this "new" ASPM gene allele frequency is approximately 50 percent of the human population. Researchers still cannot reveal the exact role of this new ASPM allele. About 10 percent of people in Europe were found to have two copies of this new ASPM allele; 50 percent have two copies of the old allele, and 40 percent of humans had one copy of each. However, no difference in intelligent quotient (IQ) was found between the different carriers of the old and new alleles.[20]

One of the leading experts in human brain evolution, Dr. Christopher Walsh, presented at the Jerusalem conference in June 2018 that the neural progenitor cells are located close to the cerebral spinal fluid (CSF). It was found also that CSF has growth-promoting signals for the neural progenitor cells. ASPM promotes the progenitor cells to become radial glial cells and start their migration to the cortex. It also disrupts the glue of progenitor cells to the CSF, provoking the progenitor cells to "decide to move" to the cerebral cortex. Defective forms of ASPM genes are described to be associated with primary microcephaly.

Microcephalin (MCPH)

Microcephalin is a gene that is expressed during fetal brain development. It is located on the eighth chromosome. One of the last genetic variants of microcephalin has been identified in humans. It is believed that this variant emerged approximately thirty-seven thousand years ago. Another human variant of the microcephalin gene arose around five thousand eight hundred years ago. Like the previous gene ASPM, a detrimental mutation in the microcephalin gene can lead to microcephaly.

[20] Zhang, J.J. *Genetics*, 65 (4), 2063–2070, Dec 2003.

The FoxP2 gene

FoxP2 stands for forkhead box protein-2. The gene is located on the short arm of the seventh chromosome. The FoxP2 gene produces the FoxP2 protein necessary for proper development of speech and language. This gene is expressed in the fetal and adult brain, heart, lungs, and gut. FoxP2 provides a prime example of protein coding sequence mutations. A comparison of the FoxP2 DNA from multiple species indicates that the human FoxP2 protein differs at only three amino acid residues from the mouse orthologue and at two residues from chimpanzees, gorillas, and rhesus macaques. These amino acid changes are functionally critical as humans' and chimpanzees' FoxP2 genes have strikingly different transcriptional targets, many of which are involved in brain development. It was found that Neanderthals and Denisovans have the same FoxP2 gene as humans. In *Scientific American Mind 2018*, Dr. Herculano-Houzel reported that we have, by far, the most cortical neurons of any other brain on earth. Human brains have the most of higher-order neurons in the cortical and especially the prefrontal area of the brain. [21]

Adam Rutherford confirmed that FOXP2 encodes a transcription factor. Transcription factors are proteins whose only function is to clamp into specific bits of DNA and promote the gene expression of that gene. According to Rutherford, we are biologically programmed for speech. Our brain is a software platform prepared for language acquisition.[22] A short communication in *Nature*

[21] *Scientific American Mind*, 2018.

[22] Adam Rutherford. Humanimal. How Homo Sapiens Became Nature's Most Paradoxical Creature, A new Evolutionary History, NY: The Experiment, LLC, 2018–2019.

explained that the FoxP2 gene acts as a "dimmer switch." FoxP2 binds to DNA and helps to determine what extent other genes are expressed as proteins. Scientists from Los Angeles also found that the human version of the FoxP2 gene increases the expression of sixty-one genes and decreases the expression of fifty-one genes compared with the chimp version of the FoxP2 protein. FoxP2 genes also affect soft tissue formation and development, linking FoxP2 to the physical side of speech and articulation. The *Nature* short communication author wrote that "language in humans did not develop from scratch." It depends on the tuning of genetic pathways present in our nonverbal ancestors rather than the emergence of completely novel mechanisms. Birds inserted with the human FoxP2 gene sing more complex songs.[23]

According to Paabo et al., engineered mice that express the human FOXP2 gene have larger dendrites. They perform better at forming synapses and running in a T-shaped maze. They also found that these mice have enhanced dopamine activity in the part of the striatum involved in forming procedures. Human FoxP2 facilitates learning that has been conducive to the emergence of speech and language in humans. [24] A FoxP2 gene mutation was found first in one South American family the 1990s. This family had three generations of patients who had trouble speaking.

---

23 Nature, 2009, 1089, online edition.

24 Science Daily, 9/15/2014.

Dear reader, if you have followed these complex genetic writings so far, I am pretty sure you will continue. This chapter is exclusively genetic and medical, but I have confidence in you. Just write out your questions. I am ready to discuss them. The first six gene mutations listed above and their roles in human brain growth have already been discussed in the scientific literature. The following twenty-two genes are seldom, if at all, studied to date. I needed to delve deep in the present literature, searching any information about their potential contributions to our brain growth and its miraculous complexity. Below are the results of my "gene delving."

The RELN gene

In humans, the reelin gene is located on the seventh chromosome producing reelin protein. According to G. J. Meyer, the reelin protein is expressed in Cajal-Retzius neurons in the brain. These neurons as well as the reelin are critical for appropriate cortical development. It was discovered that the vital brain-derived neurotrophic factor regulates reelin expression in Cajal-Retzius neurons. These neurons degenerate and undergo cell death when cortical migration is completed.[25] Besides this important role in early development, it is now also known that reelin "continues to function" in the adult brain. It modulates synaptic plasticity by enhancing the induction of long-term potentiation and memory. It also stimulates dendrite and dendritic spine development and regulates the continuing migration of neuroblasts generated in adult

[25] Meyer, G.J., *Anat.*, 2010.

neurogenesis sites like subventricular and subgranular zones of the brain.

HARs complexes

Human accelerated regions (HARs) complexes are non-coding RNA genes. There are more than three thousand HARs throughout the human genome. They were first described in 2006 as a set of forty-nine segments in the human genome, localized on the twentieth chromosome. HARs are conserved throughout vertebrate evolution, but in vertebrae, they are strikingly different from human HARs complexes. HAR1 throughout HAR49 are named according to their degree of difference between humans and chimpanzees. HAR1 displays the most considerable difference between humans and chimpanzees, whereas HAR49 has the slightest degree of difference. The gene HAR1 is expressed in neurons of the developing human neocortex. Evolutionary analysis revealed that, although this gene was only 118 nucleotide pairs in length, it contained eighteen changes in the human lineage since the divergence from chimpanzees—more than ten times the expected neutral rate. These many changes over such a short evolutionary period are truly striking. The HAR2 gene includes the HACNS1 gene enhancer that may have contributed to the evolution of the uniquely opposable human thumb. HAR2 could also have contributed to the modifications of the ankle and foot that allow humans to walk on two legs. Evidence to date shows that of the 110,000 gene enhancers identified in the human genome HARs, HAR2 has undergone the most changes during human evolution following the split with the ancestors of chimpanzees.[26]

Novel open reading frame mutations in HARs reveal new leads to understanding schizophrenia and bipolar disorder. The recent study of Eradi and colleagues is the first one of HARs mutations in these disorders, and the scientific world expects more discoveries in this area of research.[27]

The MYH16 gene (myosin heavy chain 16)

The MYH16 gene encodes a myosin heavy-chain protein present in skeletal muscles. In nonhuman primates, MYH16 is expressed exclusively in muscles of the head, including those involved in mastication (chewing food until crushed or grounded). There is also an intriguing example of gene loss that might have played a part in the emergence of the larger human brain. In humans, however, a frameshift mutation resulted in the loss of function of this gene.

In research carried out in 2014, it was found that this MYH16 gene mutation in hominids happened 1.4 million years ago. According to it, this mutation made our masseter muscle three times weaker than those of chimps. This led to the suggestion that the loss of MYH16 was partly responsible for freeing the hominid cranium from the strong masticatory muscle constraints and enabling it to expand to accommodate increased brain size. Thus, the loss of MYH16 might have come about through the relaxation of functional constraints on masticatory

[26] Levchenko, A. et al: Genome Biol. Evol., 166–188, 10, (1), January 2018.

[27] See Chaitanya Erady, Krishna Amin, Temiloluwa O. A. E. Onilogbo, Jakub Tomasik, Rebekah Jukes-Jones, Yagnesh Umrania, Sabine Bahn & Sudhakaran Prabakaran. Molecular Psychiatry, 2021.

muscles, coupled with positive selection for increased brain size.[28].

Kevin Laland suspected that the MYH16 gene deletion in the hominins coincides with the appearance of fire and cooking. This single mutation may have helped us to have larger skulls and more brain growth, especially in the frontal lobes. Other primates continued to have three to seven times stronger jaw muscles that restricted the growth of their skulls. A richer diet, especially in meat and fish, would have opened the door to further brain expansion. Eating cooked food led to the shrinking of our gut and expansion of our brain, suggested primatologist Richard Wrangham at Harvard University.[29]

> I will agree with you, dear reader, that we observe a paradox here. Our masseter muscles became three to seven times weaker, significantly decreasing our fighting capabilities. Instead of perishing, our species found a way, through its enlarged brain, to better use fire for cooking food and better socializing. As we continued growing a bigger and more complex brain, we continued to thrive and multiply.

CAG triplets

The average number of cytosine-adenine-guanine (CAG) triplets in humans is seventeen. The CAG gene repeats are localized on the fourth chromosome. It was established that people with CAG-30 triplets have enhanced intellect, but if this number increases over thirty-six,

---

[28] Hansell Stedman et al., Nature, 428, 415–418.

[29] Kevin Laland, Darwin's Unfinished Symphony. How Culture Made the Human Mind, Princeton. NJ: Princeton University Press. 2018,

one can develop Huntington's disease. If the number of CAG triplets increases, especially above seventy to eighty, Huntington's disease is more severe and occurs earlier in life. There are no studies explaining the role of the CAG triplets' number in human brain growth and evolution.

Huntington's disease is one of the thirty or more triplet diseases known to humankind. The disease is named after the American physician George Huntington, who contributed a classic clinical description of the disease. Dr. Huntington described an unknown disease in his only paper in 1872, at age twenty-two, a year after receiving his medical degree from Columbia. Dr. Huntington worked in his family-owned medical practice. His father and grandfather were also physicians. They all observed a large family with strange involuntary movements and progressive dementia in East Hampton on Long Island. Dr. George Huntington observed and described the autosomal dominant transmission in multiple generations of this large family from East Hampton. What a brain and knowledge he had for 1872! Dr. George Huntington died at age sixty-six from pneumonia under the care of one of his sons, also a physician.

RNF213 (ring finger protein 213)

The RNF213 gene is located on the seventeenth chromosome and produces RNF213 protein. This gene is believed to be a lucky accident in primate evolution that occurred ten to fifteen million years ago. RNF 213 protein is involved in forming new vessels and improves blood supply to the primate brain. It is believed that RNF213 may have played a role in facilitating the evolution of larger brains in primates.

It was shown recently in fifteen Japanese families that a mutation of the RNF213 gene causes the relatively rare but dangerous neurological condition called moyamoya disease. *Moyamoya* in Japanese means "puff of smoke." Japanese authors had observed that these patients develop gradual obliteration of a big brain vessel, usually one of the carotid arteries. Because this process of large vessel obliteration is relatively slow, the brain develops additional small capillaries to compensate for this loss of blood circulation. These small vessels look like "puffs of smoke" on the brain angiography or brain MRA. The name came because of this small artery appearance of "puff of smoke." Luckily, moyamoya is very rare, except in Japan. In my close to fifty years neurological practice, I have had the chance to diagnose and participate in the treatment of only one patient with moyamoya disease. Usually, these patients need highly qualified neurosurgical intervention. The neurosurgery procedure consists of precise connection of the external carotid artery branches supplying blood to the temporal muscles to the branches of the internal carotid arteries branches supplying the under-perfused brain. This provides additional blood supply to the poorly vascularized underlying brain tissues. Additional treatment with medicaments is also necessary. The patient with moyamoya whom I followed for three years had good reperfusion of the brain and a good outcome, thanks to the very specialized vascular surgery performed by a highly skilled neurosurgery team in University Hospital in San Antonio, as well as adequate medical treatment.

X-lined MAO-A (monoamine oxidase-A gene)

The MAO-A gene is located on the X-chromosome. It encodes a mitochondrial enzyme that catabolizes several neurotransmitters, including dopamine, serotonin, and norepinephrine. This gene is also called the warrior gene because its deficiency leads to high-level mediators such as adrenaline and a high level of aggressive behavior. In addition, MAO-A activity level increases with age, which makes older people less aggressive.

Compared with other primates, only one human-exclusive non-conservative change is present in the gene: Glu151Lys.[30] This is an example of positive selection. In contrast to humans, apes show diversity levels statistically compatible with neutrality. The observation that the closest phylogenetic relatives of humans do not show traces of positive selection for the MAO-A gene, whereas humans do, suggests the presence of a recent exclusive episode of positive selection in humans. According to McDermott et al., the MAO-A warrior gene may have played a significant role in the evolution of social behavior and makes one more likely to engage in "physical aggression."[31]

After further detailed research in the literature, I found fourteen more genes potentially capable of multiplying our neurons or improving the supporting glia that can be involved in our human brain growth and development. I am afraid that elaborating on them would involve too much genetic knowledge for now. We will wait for more research in this area. Here are only the names of these genes waiting for future research:

[30] Aida Andres, "Human Genetics," 115 (5):377–86, 2004.

[31] *PNAS*, February 17, 106(7) 2118–2123, 2009.

SHH (sonic hedgehog)
AHI1 gene (Abelson helper integration site 1)
GLUD2 (glutamate dehydrogenase-2 gene)
Neurexin and neuroligin
NFIX (nuclear factor one X)[32]
Vacuolar protein genes[33]
FGFs (fibroblast growth factors) and FGFRs (fibroblast growth factor receptors)
Homeobox protein engrailed-1 (En1)
The En2 gene (engrailed-2) gene[34]
ULK4 (uncoordinated 51-like ligase) gene[35]
SOD1 (superoxide dismutase 1) gene
ADGRG1 gene
FADs (fatty acid desaturases) gene
CMAH gene (cytidine-monophosphate-N-acetylneuraminic acid hydroxylase gene)[36]
NOVA 1 (neuro-specific RNA-binding protein)
miR-941 (microRNA-941)[37]

---

32 Reported in the Jerusalem Conference 2018, lecture 2.

33 See Laland, loc. cit.

34 Barbara Kuemerle et al. "The Mouse Engrailed Genes: a Window into Autism," Behav Brain Res., 176(1); 121–32, Jan. 10; 2007.

35 Matt Garland. *Self-Decode*, 2019.

36 Gardien, July 2019.

37 See the article by Hai Yang Hu, Liu He, Kseniya Fominykh, Zheng Yan, Song Guo, Xiaoyu Zhang, Martin S. Taylor, Lin Tang, Jie Li, Jianmei Liu, Wen Wang, Haijing Yu, Philipp Khaitovich. "Evolution of the human-specific microRNA miR-941," Nature Communications, 2012; 3: 1145 DOI: 10.1038/ncomms2146.

Apparently, we humans are the winners of the genetic jackpot because these gene mutations have helped our brain grow to be three times larger than the nearest primate during the last six to seven million years. At the same time, some of these "brain-helpful" gene mutations became "easy breakable points" for a multitude of neurological and psychiatric uniquely human disorders. At the beginning of 2022, we are still in the process of collecting scientific genetic data. We need scientists and multidisciplinary researchers to confront the complexity of the brain formation and ways to use our potential for better preventing, diagnosing, and curing our diseases.

Dear reader, if you have read through all these pages, I can appreciate your exemplary stoicism. The rest of the book will not be so complex. All twenty-eight genes, twelve of which I described in this chapter, are poorly known even by the medical community. The most important thing to remember from this genetic chapter is that we humans, like all other living animals, have been "bombarded" with an unknown large number of gene mutations—for better or for worse.

"The brain is a citadel that cannot be taken by direct assault."
—Charles Darwin

CHAPTER III

The Environment and Its Role in Human Brain Growth and Evolution

"Genes load the gun; the environment pulls the trigger."
—Unknown

"Genetics is not a destiny but a guidebook which can be modified by food and environment."
—Dr. Sudha Seshadri, Director of Biggs Alzheimer Institute, San Antonio, Texas

Living creatures evolve, which enables them to adapt to constantly changing environments. This process requires two adaptation mechanisms: species-level genetic changes and individual-level changes. The changes explained in this chapter occur during species' development and rely on adaptive mechanisms that require constant epigenetic modifications. The science that studies these modifications is called epigenetics.

3.1 Epigenetics

The term "epigenetics" was introduced in 1942 by embryologist Conrad Waddington. The first human disease to be linked to epigenetics was colorectal cancer in 1983. The pioneers of epigenetic studies were neurobiologists Michael Meaney and Moshe Szyf at McGill University. In 2005, they showed that rat pups of inactive mothers grow up to be very anxious. On the contrary, the baby rats of the active mothers, who regularly lick their pups, grew up to be calmer.[38]

Bill Sullivan described the two main mechanisms of epigenetic influence on gene expression: DNA methylation and histone acetylation. The first mechanism is the methylation of the cytosine in the DNA chain, which turns the methylated gene "off." The methylation slows the "traffic" in gene expression. The methylation and demethylation of nucleoside cytosine sites are known to be the primary source of epigenetic modifications in regulatory DNA regions. The author of this book makes it simple to remember: the methyl groups, chemically represented as CH3, "are bumps on the road that slow the traffic in the gene expression" or "methylation mutes the gene." The second mechanism of gene regulation involves acetylation of a group of proteins called histones. Histones are wrapped around the genes like a piece of thread. Adding acetyl to a histone protein "tunes" the gene expression rather than switches them "on" and "off" as done by the methylation of the DNA cytosine. The acetylation of histones acts as a dimmer switch rather than an "on" and "off" switch. Both mechanisms described

[38] "Dialogues Clin." Neurosci. 2005; 7(2), 103–23).

above change the gene expression without changing the DNA sequence itself. These changes are called epigenetic, meaning in Greek "beyond the gene." Epigenetic modifications allow the environment to send a message to our genes, altering how they work for not only for us but also for our children and grandchildren. As the famous botanist Luther Burbank noted, "Heredity is nothing but a stored environment." This can be a great advantage to us and our children because the rapid changes of gene expression allow quick adaptation to environmental conditions.[39]

Several years ago, *Life Science* published a comprehensive article on the above observation, saying that basically DNA methylation and demethylation is a normal mechanism used by cells to control gene expression. Methylation of the cytosine silences the gene; demethylation makes the gene active. In addition, the *Life Science* journal article reports that alteration of DNA methylation can also be initiated and influenced by the following:

Oxidative stress – While oxidative stress does occur naturally and contributes to the aging process, oxidative states significantly influence a wide range of biological and molecular processes and functions.

Heavy metal exposure – In addition to mercury, other heavy metals that are classified as human carcinogens, including arsenic, lead, and cadmium, can influence our genome epigenetically. For example, early exposure to lead in

[39] Bill Sullivan. *Pleased to Meet Me. Genes, Germs, and the Curious Forces that Make Us Who We Are.* National Geographic, 2019.

children alters DNA methylation patterns in genes linked to brain development and neurological disorders.

Lifestyle factors are of crucial importance in epigenetic modifications. Stress, unresolved past emotional trauma, and other lifestyle factors such as unhealthy diet, smoking, lack of exercise, and excessive alcohol consumption are additional factors to be considered.

Stress is linked to everything, from the common cold to depression and heart disease. Stress also causes the body to lose its ability to regulate inflammation. The stress hormone cortisol usually plays a role in regulating immune cells. Still, when stress is prolonged and becomes chronic, "those cells become insensitive to cortisol and inflammation goes unchecked."[40]

Muscle activity – Professor Herman Pontzer reported that human legs are longer and have bigger muscles, with a greater proportion of "slow twitch" than the legs of apes. The author observed that the proportion of our legs' "slow twitch muscles" is 70 percent (versus 30 percent "fast twitch muscles") and is higher than any other nonhuman primates. These "slow-twitch muscles" primarily use their mitochondria as energy provider and secure our dominance as endurance, long-distance runners. We also have more red blood cells to carry oxygen to working muscles. Endurance exercises decrease chronic inflammation and lower cortisol levels. At the same time, exercising improves the effectiveness of our immune system. Muscles produce enzymes that help to clean

[40] Dr. Mark Hyman, The Doctor's Pharmacy minisode # 42, "How to stop Stress controlling your Life," video-interview.

fat from circulating blood. Exercise increases multiple neurotrophic factors, the most known BDNF (brain-derived neurotrophic factor), promoting neurogenesis and brain growth.[41] As John Medina emphasized, physical activity stimulates blood vessels to produce nitric oxide, the blood regulatory molecule. As blood flow improves, the body forms new vessels. BDNF is connected to the memory of past events. Dr. Medina wrote, "The good microbiome enhances expression of BDNF in the hippocampus, regular exercise, too." Exercise also changes gene expression by reprogramming one's genome.[42]

Diet – The detailed *Life Science* article also summarized the study of Romain Barres and colleagues on DNA methylation patterns in sperm samples from lean and obese men. In this study, Barres's team from Denmark reported more than nine thousand genes methylated differently in the sperm from lean versus obese men. These alterations in gene expression could lead to permanent changes in how the brain is wired for appetite.

The role of sleep will be discussed next.

---

[41] Herman Pontzer. "Economy and Endurance in Human Evolution. Current Biology, volume 27, issue 12, June 19, 2017.

[42] John Medina, *Brain Rules in 12 Principles for Surviving and Thriving at Work, Home and School,* Edmonds, WA: Pear Press. 2008.

Dear reader, before even touching on the issue of the very important role of sleep on epigenetics and the development of our brain, I would like to give my comments on the connection of melatonin, sleep, and health. San Antonio University is known worldwide because of Professor Russel Reiter. At the age of eighty-seven, Dr. Reiter continues his intensive research on melatonin, which he begun more than sixty years ago. According to Dr. Reiter and the leading experts in the field, melatonin is not a simple hormone but a universal antioxidant and anti-inflammatory agent. On May 30, 2021, Dr. Reiter was the honored guest at our La Madeleine Sunday group. For more than three hours, he discussed with us his findings on melatonin in health and disease. Below, on the picture, Dr. Reiter is the second sitting gentleman from the right.

Melatonin function is too multifaceted and complex to be discussed in this book. Therefore, I will focus mainly on the melatonin anti-inflammatory function. We had additional stimulus to do so because we are still dealing with the coronavirus pandemic. In the past two decades, a few medics clinically observed that melatonin, used in "higher than sleep" doses, also has an anti-inflammatory effect. This was first observed in some Ebola-infected patients by our colleague and friend Dr. Richard Neel. Dr. Neel, sitting on the right of Dr. Reiter, had these observations when serving as a US military physician, but no studies had been performed at that time. As I write these lines, we continue to be in the COVID-19 pandemic. Very few studies have been written about melatonin and its anti-inflammatory effect in COVID-19 infected patients.

In March 2020, Dr. Richard Neel started recommending higher doses of melatonin as an anti-inflammatory and safe medication for patients with initial COVID-19 symptoms. His first interview on this subject was published in the *San Antonio Express-News* on April 16, 2020. Hundreds of patients from all over the United States who read this interview telephoned Dr. Neel, and he generously guided their melatonin and other medication treatments according to their COVID-19 symptoms. He also treated many patients in his emergency rooms, "Little Alsace," in Castroville and Hondo, small towns near San Antonio. Dr. Neel recommends treatment with 50–100 mg, usually 1 mg/kg of oral or sublingual melatonin a day, split in four doses for a period of seven days, starting from the onset of the COVID-19 symptoms.

In November, Professor Reiter and Dr. Neel were asked for Zoom lectures and interviews attended by more than five hundred Canadian medical practitioners. Dr. Neel, his medical partners, and their helpful staff are still too busy with COVID-19 and many other patients in their clinics. Unfortunately, melatonin is a cheap, generally available over-the-counter medication, and it does not stimulate any placebo-control studies. There was, to my knowledge, only one small positive placebo-control study from the Philippines about applying oral and intravenous melatonin in COVID-19 patients. Canadian colleagues showed real interest in higher doses of melatonin use in COVID-19 and other viral and infectious diseases. We, neurologists, and sleep specialists know that extra melatonin does not suppress our own melatonin production.

Interestingly, I vividly remember one lecture by Dr. Reiter for one of our Friday Neuroscience Grand Rounds in 2008. Dr. Reiter emphasized that only 5 percent of our melatonin is produced by the pineal gland. The remaining 95 percent is produced by our mitochondria, which they use completely without sharing a single bit. The mitochondria need all their melatonin to neutralize the free radicals they produce. In another lecture in 2019 in Brazil, Dr. Reiter underscored that melatonin production in people after the age of seventy is twenty times less than in children. Dr. Reiter also recommends the use of blue light–blocking glasses in the evenings. It is well-known that blue light suppresses our pineal gland production of melatonin. Blue light is emitted from all electrically powered devices, especially TV and computer screens. For the last three years, I have strictly followed his advice.

Why not? These blue light–blocking glasses cost only fifteen dollars, and they increase our melatonin production. There is good news for all romantics: moonlight does not emit blue light since the moon merely reflects sunlight. Melatonin is not suppressed by moonlight.

3.2 The role of sleep in human brain growth and evolution

The importance of a good, restorative sleep for human health has been known for many centuries. However, despite much has been written, very little is known about the role of human sleep benefit for our brain growth and evolution. Humans average seven to eight hours of sleep, great apes eight to nine hours. We spend 10 percent more time in REM sleep than greater apes. Gibbons sleep in an upright position, hanging on the branches of different trees. They do not build sleeping nests like other great apes. Human sleep is a little shorter, but deeper.

In one of the few comparative literature sources on sleep in animals and humans, Dr. R. Douglas Fields wrote that nonhuman primates have only three to four REM sleep cycles per night; we humans have, in general, five to six REM sleep cycles. Other than his book, we could not find more quantitative or qualitative studies on the role of sleep structure *per se* on human brain development compared with nonhuman primates. Dr. Fields continued, "Astroglia influences gamma-waves. Learning is dependent on gamma-wave activity." Sleep is ultimately involved with learning. [43]

John Medina suspected that humans appear to replay their daily learning experiences during slow wave sleep. It appears that one of the reasons we need to sleep is so that we can learn and rehearse our experience from the previous day. However, according to the Centers for Disease Control and Prevention (CDC), more than 25 percent of people in the United States often fail to meet sleep recommendations, and around 50 to 70 million American adults have some form of sleep disorder. The estimate is that the effect of sleep deprivation costs approximately one hundred billion dollars a year to US businesses. [44]

I learned the basic of sleep medicine with my Bulgarian teachers. Some of them studied under the well-known professor Michel Jouvet in France. Dr. Jouvet was the first to establish in the 1960s that muscles in cats during their REM sleep remain completely flaccid, despite their recorded EEG activity during this time, somewhat comparable to their waking EEG. For this reason, Dr. Jouvet named the REM sleep "paradoxical sleep." As we know, REM sleep is named first in 1953 by Eugene Azerinsky, a PhD student, and his adviser, Dr. Nathaniel Kleitman. The student Azerinsky spent long nights to observe and register the eye movements and EEG of sleeping infants.

[43] R. Douglas Fields, Electric Brain, How the Science of Brainwaves Reads Waves, Tells Us How We Learn and Helps Us Change for the Better, Dallas, TX: BenBella Books, 2020.

[44] John Medina, Brain Rules.12 Principles for Surviving and Thriving at Work, Home and School. Edmonds, WA: Pear Press. 2008.

He and his adviser, Dr. Dement, demonstrated that "rapid eye movements" (REM) are correlated with dreaming and a general increase in brain activity seen in EEG. After this pioneering work, scientists understood for the first time in centuries that sleep is not a uniform and passive event but has at least two components. Thereafter, the names REM sleep and slow-wave sleep started to be used routinely.

Later, in 1992–1995, I used my "Bulgarian sleep medicine experience" to perform the first sleep studies on a patient suspected to have a narcolepsy in the clinic of Dr. Todorov in Tuscaloosa, Alabama. Narcolepsy is a relatively rare neurological disorder expressed with bouts of increased sleepiness and fatigue during the day. Sometimes these episodes are associated with sudden loss of muscle tone and occasional falling called cataplexy. The neurophysiological study for testing narcolepsy is called MSLT (multiple sleep latency test). This test is performed during daytime, after night sleep evaluation to make sure that the patient had sufficient sleep the night before. Dressed comfortably, the patient is put in a recliner and instructed to relax and try to sleep every two hours. Between these naps, the patient needs to be kept awake. We usually collect four or five naps. The test is positive for narcolepsy if the patient has two or more naps with REM sleep registered in the first twenty minutes of sleep. Usually, the first REM sleep is registered in 80–120 minutes of superficial and slow-wave sleep.

An interesting continuation of the narcolepsy story is that in 1998, when preparing for my medical exams, I had the chance to meet at a course and have a lunch with Dr. Emanuel Mignot. At that point, few people knew him. He worked in the famous Stanford Sleep Clinic with another known sleep researcher, Professor Christian Guilleminault. Dr. Mignot told me about his research on Doberman dogs with narcolepsy. Two years later, the entire medical world knew about Dr. Mignot and his discovery of the defective protein in narcoleptic Doberman dogs. He called it hypocretin.

Many inventors claim their ideas came in sleep or soon after sleep. One of them is August Kekule, the inventor of the benzene ring (a basic component of many organic compounds). He recounted a dream in the winter of 1861–1862 while dozing in front of a fire in Ghent, Belgium, where he was a professor of chemistry. Kekule dreamed of a self-devouring snake, which led him to the idea of the benzene ring.[45] Dmitri Mendeleev was another world-known discoverer who shared that his idea about the periodic table of the chemical elements came in his sleep.

[45] The "self-devouring" snake, the serpent swallowing its tail, is an ancient Greek symbol signifying the eternity of the universe. Kekule was undoubtedly familiar with the symbol and its significance.

The renowned Google co-owner Larry Page described a similar experience in 1995 at age twenty-three. At that time, Larry had a dream that a World Wide Web search engine could rank links based on how often they were linked to other pages. The following day, he made a small initial capital contribution and shared his idea with his classmates, "computer gurus" Sergey Brin and a Bulgarian student. The Bulgarian student was skeptical and refused to join. Sergey Brin accepted. This story was told to us by my oldest daughter, Svetla, who has been working for Google for the last several years. Later, she invited us to attend Parents' Google Day, and we heard part of the story from the founders of Google themselves. We also heard a continuation of the story. In 1999, Larry and Sergey decided to change the primary name of their company called Backrub. A new name was suggested by one of their collaborators. The name was "Googol," which means the enormous number equal to ten followed by one hundred zeros. For some reason, Larry spelled this as "Google," and the name stuck perfectly.

3.3 The role of agriculture and animal domestication in human civilization and development

According to Richard Dawkins and Yan Wong (evolutionary biologists and authors of the book *The Ancestor's Tales*), the "invention" of agriculture encouraged specialization in human societies. Without our domesticated animals and plants, human civilization as we know it would not exist.[46] According to Richard

[46] Richard Dawkins and Yan Wong, The Ancestor's Tales. A pilgrimage to the Dawn of Evolution, Boston: Houghton Mifflin, 2016.

Francis, domestication represents an accelerated form of evolution, including human self-domestication from hominids to hominins. Our evolutionary changes are especially manifested in the human prefrontal cortex and in another part of the human brain called the caudate nucleus. These accelerated gene changes in humans are not seen in chimps' brains when the two are compared.[47]

> Scientists believe now that the first domesticated animals were goats around ten thousand years ago. It is important to remember this timing because multibillionaire Jeff Bezos invested forty-two million dollars to build a clock in a cave at a six-thousand-foot altitude in his newly acquired ranchland in Texas and New Mexico. This mechanical and electrical miracle will measure time accurately for the next ten thousand years. When we see this marvel, we may have a better appreciation of the last ten thousand years of agriculture and domestication of animals, helping us to survive and reproduce and soon to reach the eight billion benchmarks.

---

[47] *Richard C. Francis. DOMESTICATED. Evolution in a Man-Made World.* NY: W. W. Norton, 2015.

The domestication of sheep, cows, pigs, and horses followed. People started to keep excess animal milk in pouches of animal skin around their belts at a body temperature of around ninety-seven degrees Fahrenheit. Shepherds, at that time, presumably seven thousand years ago, observed that animal milk took on a harder consistency. In the ancient lands of Thrace, today my homeland, Bulgaria, they called it "yo-gurt," translated by some historians from the old Thracian language as "hard-milk." Almost nobody knows today that when we shop in the grocery store, we use a seven-thousand-year-old word "yo-gurt" from an obsolete Thracian and probably Dacian language. Their culture of making yogurt is still preserved in the Balkans on both sides of the Danube where people discovered over time how to concentrate the fermented milk and make cheese. People then and the following several millennia probably began producing a more active lactase, the enzyme that helped adults digest other animals' milk.[48] As we know, only babies and young infants need the enzyme lactase to digest their mother's milk. Once adults were producing more lactase, they added more milk products to their diet, providing the early "pastoral" people with an additional source of valuable food, especially for the brain. Several more millennia passed—until the early 1900s—when scientists began to understand how certain bacteria in "yogurt" and cheese help our "good microbiome." Later, in this book, I will discuss probiotics, the microbiome, and their role in human health and development.

[48] Lactase is the enzyme that helps break down *lactose*, a type of sugar found in milk that is difficult for some adults to digest.

3.4 The complex question of human gender identity and our brain

> The question of genders and gender identities is discussed in hundreds of books and probably a thousand articles. In the following short paragraphs, I will discuss some biological factors regarding the "hot topic" of gender identification and same-sex attraction. I think this intricate question has and will be important for our societal evolution. I also wish to popularize the biological aspect of the gender question.

Presently, the biology of male homosexuality is better understood than the female one. It is known that male babies' brains, exposed as embryos to a mother's low level of serum testosterone, sometimes fail to develop as brains of heterosexual males. These suppositions are only suggested in the literature with no clear scientific evidence. Another biological explanation for male-to-male sexual attraction is that the antibodies produced by the pregnant mother bind to proteins on the surface of the cells of the fetal anterior hypothalamus, the area associated with sexual orientation. This possibility is also discussed by Graham Lawton and Jeremy Webb.[49] It is unclear how many other factors play a role in the sexual orientation of teenage boys and young men, and the question is beyond the main scope of this book.

The biology behind female homosexual behavior is probably even more multifactorial. However, let me present here one of the little-known and largely ignored

[49] "How to be Human. The Ultimate Guide to Your Amazing Existence," London: John Murray Publishers, 2017.

biological fact that can be one of the drivers for female homosexuality. At puberty, at age eleven or twelve, our pituitary gland starts producing many gonadotropin hormones. Amazingly, the pituitary hormones provoke the same initial response from the testes and ovaries in young boys and girls; they spike the production of testosterone. In boys, the testosterone stays at high level. In most girls, typically, the testosterone is quickly converted to estrogen facilitated typically by a very active enzyme in females. This enzyme is well-known and is called aromatase. Consequently, the level of testosterone in most young women drops quickly and stays at a low level during their reproductive lives. Usually, the newly formed high levels of estrogens, influenced by an active aromatase enzyme, help tune the brain cells to respond to estrogens and develop feminine behavior. This estrogen hormone action on the brain helps build the sexual behavior that will further secure a desire for heterosexual contacts and procreation. In girls with low levels of the enzyme aromatase, this conversion of testosterone to estrogen is inhibited, and their levels of testosterone may stay high, promoting male behavior. At this point, there is not active research in this area.

I found a few articles from the 1980s discussing the level of testosterone in girls who look for loving partners among their own sex. I could not find any studies regarding the level of the enzyme aromatase in females of this age. I may presume that low aromatase levels in some girls result in higher testosterone and lower estrogen concentrations.

Such a disbalance could be one of the multiple ways the brain in some adolescent girls develops same-sex attraction. Presently, we need far more objective investigations in this area rather than giving unguided opinions on this very sensitive subject. It is better to present more research and less prejudice. Science should connect specialists in different areas, such as genetics, obstetrics and gynecology, endocrinology, neurology, psychiatry, and psychology, to start serious multidisciplinary research of sexual differentiation, especially before and during puberty. We need to know far more about all androgens and estrogens and their enzymes and carrier globulins before we can pretend to know something about the whole complexity of sexual orientation and gender identification. It could be a challenging but not impossible task.

3.5 Viruses and their role in human evolution

Viruses are a string of RNA or DNA packaged in a protein coat. Their first emergence is traced to some three billion years ago. Currently, science estimates that there are more than eight hundred thousand types of viruses, with only around six hundred known to be pathological to humans. One of them, coronavirus, became very notorious in the last few years. The name "corona" depicts the structure surface of the virus, a spherical corpus with spikes looking like a crown, *corona* in Latin. Coronavirus is composed of RNA containing around thirty thousand nucleic acids forming twenty-nine genes. One of these genes, known as a spike protein, attaches the virus to our cells' membranes. Coronavirus underwent its first major known mutation in 2002. This mutation made the virus more attachable to the membrane and with higher penetrance in the cells. The disease caused by this virus was called SARS (sudden acute respiratory syndrome), which first appeared in China. The Chinese government and medical community could contain the infection very well at that time. As a result, approximately only seven hundred people were infected, and four hundred of them died.

At the end of 2019, an even more contagious and dangerous coronavirus spreading infection paralyzed almost all the world. On January 11, 2020, Chinese scientists announced that they had deciphered the thirty thousand nucleotides of the new coronavirus. They named this mutated coronavirus SARS-CoV-2. The Chinese government and medical community could not contain the infection this time, which started in the large city of Hunan. In October of 2020, scientists found the exact site of the new mutation, located on the tip of the coronavirus spike protein. This mutation allowed the newly mutated coronavirus to attach ten times more firmly to a specific protein on the surface of the human cell membrane, further penetrating the cells through the ACE2 membrane receptors. This has led to a global epidemic with more than 1.8 million deaths reported at the end of 2020. This mutation gives the virus a strong membrane adherence, allowing the coronaviruses to enter our cells in huge quantities compared with previous coronaviruses. In January 2021, at least two new variants of the mutated virus were found. They were 50 to 70 percent more contagious than the original variant mutated in 2019.

In mid-December, vaccination programs began first in the United Kingdom and the USA with mRNA based on coated spike protein. In Russia, researchers used the spike protein attached to two human-modified adenovirus carriers. Some companies such as Johnson and Johnson used similar technology, but with only one shot and one type of adenovirus. China developed a "dead coronavirus" vaccine. A similar method is used largely by pharmaceutical companies in India.

The above-written paragraph is an object of study of infectious diseases, epidemiology, and other medical specialties. I just wanted to popularize common facts on the pandemic we are living through. I now return to focus on the role of viruses in our brain growth and human evolution.

In a recent review in *Science Focus Magazine*, Kate Arney discussed the extraordinary role of viruses in shaping human evolution. In her detailed article, Arney wrote, "Coronavirus, Zika, Ebola, flu, even the boring old common cold—we're all familiar with the viruses that plague humanity. But while we know they make us sick; it may be surprising to discover that, over millions of years, we've managed to harness and domesticate these crafty invaders." Arney continued, "Over millions of years, these viral genetic sequences randomly mutate and change, losing their ability to break free from their host cells. Trapped inside the genome, some of these 'endogenous' retroviruses can still jump around while others are stuck forever where they last landed. And if any of these events happens in the germ cells that make eggs and sperm, then they will be passed down the generations and eventually become a permanent part of an organism's genome." It is scientifically accepted now that around half of the human genome can be traced back to long-dead viruses or similar "jumping genes," known collectively as transposable elements or transposons. And it turns out that as well as being our genetic enemies, some of the viruses embedded in our genome have become our helpers. Many of these old viruses became our friends for better adapting in the continuously changing environment.

Here is another formidable example taken from the remarkable Kate Arney article cited below. The name of the protein is *syncytin.* This 290-million-year-old protein-derived virus, poorly known by the medical specialists and the general public, was reported to be involved in mammalian placental formation. Syncytin is a placental protein that attaches the growing embryo to the uterus. The name "syncytia" given by scientists sticks well because it makes a molecule that fuses placental cells together, creating a special layer of tissue known as a syncytium. Arney wrote, "Curiously, syncytin looks a lot like a gene from a retrovirus. Clearly, all mammalian species were infected by these particular viruses millions of years ago. Over time, those viruses were harnessed to play a key role in placental growth, making them a permanent fixture in our genome."

Arney also described the scientific work of Jason Shepherd, who found another virus-gene-friend, *the ARC gene*. It is scientifically reported that Dr. Shephard found the ARC gene accidentally. When Dr. Shepard and his lab were trying to purify the gene, they observed that the liquid was not coming out of the gel. When Dr. Shepherd reviewed the clot on the electron microscope, he noticed multiple big hollow spheres clogging the gel filter. When he analyzed the ARC capsule, he concluded that the capsule was identical to that of the HIV virus. They named this gene and its product protein "activity-regulated cytoskeleton-associated protein" or, abbreviated, ARC. The ARC gene was found only in land-living animals. Consequently, Dr. Shephard and his team and his researchers believe that the ARC gene was in fact a virus "domesticated" by the genome of the common ancestor of land-living animals some three hundred seventy-five million years ago. Dr. Shepherd called this "our biological gift. Today we already know that our memory is based mainly on the ARC gene."[50]

Neil Shubin gave another gene example of a virus turned "from a foe into a friend" that may have shaped our species. This gene is named PRODH1 (the abbreviated form of proline dehydrogenase 1). The PRODH1 gene is located on the twenty-second chromosome. PRODH1 is a mitochondrial enzyme that is involved in the breakdown of the amino acid proline, transforming it into the essential neurotransmitter for the central nervous system (CNS)—glutamate. This gene is found in our brain cells, particularly in the hippocampus. Proline will degrade to glutamate and alpha ketoglutarate, ultimately providing more energy in the form of ATP. Chimpanzees also have a version of the PRODH gene, but it is not nearly so active in their brains. Maria Sunstova and her coworkers had investigated the brains of seven humans and six chimps. They found a clear upregulation of the PRODH gene in humans compared

[50] Kate Arney. " Viruses. How They Shaped Human Evolution," Science Focus Magazine, March 19, 2020.

with chimps. Today, faults in PRODH are thought to be involved in certain brain disorders, such as schizophrenia. The above findings are described by Neil Shubin in his book cited below. He maintains that 70 percent of our genome is made by "jumping genes."[51]

Research has found that jumping genes can carry beneficial mutations that can do dramatically new things. The no-longer-jumping gene was put to work where it landed. Fortunately, though it tends to duplicate itself indefinitely, DNA has a mechanism to stop the multiplication of gene copies. Jumping genes exist to make copies of themselves like cancer cells do, but opposite to that, they are another example of "viruses beneficial for our evolution."[52]

According to Dr. Mathew Scott, evolution is deeply affected by jumping genes. Many of them have a propensity to "jump" because they possess the ability to both cut off and paste genes. The jumping gene also has the tendency to multiply, sometimes a million times. It is believed that jumping genes are seen more often in primates because the primate jumping genes more often have transposases (specific enzymes that permit the cut-and-paste mechanism). Virtually, viruses are complex transposons preserved by a protein capsule. The jumping genes can land on regulatory genes and cause cell overgrowth too. Jumping genes were long ignored by traditional science.[53]

---

[51] Neil Shubin, Some Assembly Required. Decoding Four Billion Years of Life, from Ancient Fossils to DNA, NY: Pantheon Books, 2019.

[52] According to Sara Reardon's report in Nature News, January 11, 2018.

[53] Mathew Scott, PhD, "Jumping Genes in Evolution and Medicine" Video lecture, 2014.

In 1930, Barbara McClintock worked on maize transposons and found that spotted corn derived from gene mutant insertions. These insertions are reversible, and the purple pigment of maize can recur. At first, scientists believed that the discovery of transposons was specific to maize and some other plants. Then, in 1960, scientists discovered that transposons exist in bacteria (E. coli) and, later in the 1970s, in the human genome and in most eukaryotes. Nowadays, it is established that approximately 50 percent of our genome is composed of transposons, although most of them are not functional like ancient fossil vestiges. The fundamental discoveries by Barbara McClintock were appreciated forty years later by awarding her alone the Nobel Prize in 1983. Because of this and other fundamental research, we now know that transposons are original residents of our genome. Transposases can find the ends of DNA segments, cleave them, and transport and paste them to another place in the chromosome or on another chromosome. Transposons cannot move much because they are heavily mutated and dead. But they can move without causing mutations. The genome needs to stop the uncontrolled replication of transposons; therefore, it uses an epigenetic mechanism interposing DNA segments that restrict the self-multiplication of transposons, thus maintaining the balance between functional genes and transposons. Without this balance, our genome would be invaded by fast replicating transposons and would die. As Barbara McClintock suspected, transposons have an enormous role in genetic diversity. This can serve two purposes: adaptation to new environments or development of a disease or death.[54]

[54] Susan Wessler, Riverside University Lecture. Deciphering the Strategies by a Very Successful Transposable Element. Video, February 23, 2021

During my years as an assistant professor in the Neurology Academy, I feel very appreciative that we had neurology/neurosurgery/psychiatry grand rounds every Wednesday at 11:30 a.m. Under the guidance of our professors, we had to discuss one case in each discipline and keep up with the achievements of our colleagues from the neurosurgery and psychiatry clinics too. This was an enormous help, especially for patients like one child with the rare viral disease called subacute sclerosing panencephalitis. This child was saved by the multidisciplinary team through intraventricular interferon application. These grand rounds helped us to study and practice in all three related medical disciplines: neurology, neurosurgery, and psychiatry. I feel fortunate to continue my education in San Antonio, where we continue to have our Friday morning neurology and neurosurgery grand rounds. Yes, I have continued to attend our university neuro grand rounds for the last twenty years, even as a neurologist in private practice. I thank my teachers and colleagues for organizing the weekly grand rounds in San Antonio, especially the heads of the neurology and neurosurgery departments, Dr. Brey, Dr. Jackson, and Dr. Floyd. Interdisciplinary discussions are extremely valuable, especially for the fast-changing science and medicine. Let us talk to each other! Let us talk to our patients too!

3.6 Evolutionary symbiosis or how we live together

According to Carrie Arnold, viruses that infected our ancestors provided the genetic foundations for many of the traits that define us. Scientists found out that the so-called junk DNA—a significant portion of which originates from symbiotic viruses—is a potent factor in the evolution of new species. New studies revealing the role of endogenous retroviruses in the more recent evolution of humans show that these snippets of junk DNA are helping to blur the boundary between human and virus. Retroviruses, however, use a slower, stealthier approach. After entering the cell, the retrovirus uses an enzyme called reverse transcriptase to turn its RNA into DNA before making its way to the nucleus. Once in the nucleus, it inserts its DNA into the host's genome. Most of the time, when a virus integrates its genome with the host's, the new hybrid genome dies when the cell and its descendants do. Sometimes, however, a virus will infect a sperm or egg cell. If fertilization occurs, the offspring will have a copy of the viral genome in every single one of its cells. It can pass the hybrid genome on to its offspring, creating what scientists call a fully endogenous retrovirus—a fancy term for a "virus that comes from within." Arnold continued, "The late University of Massachusetts researcher, Dr. Lynn Margulis, believed that cooperation also played a role. Beginning in the late 1960s, Margulis argued that our cells contained symbiotic bacteria known as mitochondria and chloroplasts, which earned room and board by either supplying energy or producing food from sunlight. For many years, Margulis's idea was ridiculed, and she struggled to find a journal that would publish her hypothesis. By the 1990s, however, enough genetic evidence had accumulated to show that Dr. Margulis was right. Symbiosis was responsible for some of the most significant evolutionary leaps in the history of the planet."[55]

---

[55] Carrie Arnold, in Nova Next, September 2016.

According to Bill Bryson, DNA exists for just one purpose—to create more DNA. The same author describes a type of "jumping gene" called scientifically Alu elements. Alu elements are about three hundred bases long. They are transposable elements that exist only in primates. More than one million Alu elements repeats are seen throughout our genome, including sometimes in the middle of important coding genes. Alu elements are responsible for the regulation of tissue-specific genes. They are also involved in the transcription of nearby genes and can sometimes change the way the gene is expressed. Alu elements are the most abundant transposable elements. They are rich in cytosine nucleotide, followed by guanine nucleotide, which makes them the most common site for methylation of the genes and epigenetic influence. Approximately 30 percent of methylation sites are in Alu elements, elevating them to an essential status in evolution in general and the development of our brain. Alu elements constitute 10 percent of our genetic material.[56]

The constant battle between pathogens and their hosts has long been recognized as a key driver of evolution. Scientists' findings suggest that viruses have driven an astonishing 30 percent of all protein adaptations since humans' divergence from chimps. The constant battle with viruses has shaped us in every aspect. All organisms have been living with viruses for billions of years. It was recently found that only in primates, the transposons can jump from one place of the chromosome to another location or another chromosome. In other species, when one gene or part of the DNA were duplicated,

[56] Bill Bryson. The Body. A Guide for Occupants. Canada: Penguin Random House, 2019.

the second part would be replicated just next to the first one. Therefore, the transposons in primates are unique because of this "unique jumping capability." This jumping capability in primates allows the duplication of the genes to be more beneficiary in the case of neuron production, but also the jumping site may become more prone to mutations and the site of possible genetic variants or diseases.[57]

Dear reader, you have climbed another summit of modern science connecting three-billion-year-old creatures, called viruses and transposons, "hopping freely" in our genome. Sometimes these viruses fight against us; other times, they incorporate and collaborate with our genome. Such a complex behavior is beyond the comprehension of any single person and beyond the topic of any single book, but together, we can start discussing the complex role of epigenetics, viruses, and other factors in our brain growth and evolution.

3.7 The role of the climate in shaping human brain growth and evolution

Our earth has been under innumerable climatic and disruptive changes that are probably known only to dedicated scientists. The exact story of how life on earth and our species survived on our highly climatically inconsistent planet can only be surmised. Below are some essential points clarifying how these constantly changeable conditions shaped our brain and behavior, turning us into

[57] "Viruses Revealed to Be a Major Driver of Human Evolution," Science Daily, July 13, 2016 with main reference to David Enard et al., "Viruses Are a Dominant Driver of Protein Adaptation in Mammals," eLife, 2016:5 DOI:10.7554/eLife.12469.

ultimate survivors and probably the big jackpot winners. A growing number of scientists believe that climate changes may have played a major role in shaping human evolution.

According to the oceanographer and paleontologist Peter B. de Menocal, the first evolutionary shake-up happened between 2.9 million and 2.4 million years ago. At that time, the famous ancestral lineage of "Lucy" (Australopithecus afarensis) became extinct, and two other quite distinctive groups appeared. One of them had hints of some modern-looking traits, including larger brains. Their owners were probably the very first members of our own genus, *Homo*. Crude stone tools appeared near these fossils. At the same time, another group emerged besides *Homo* that looked different: a stoutly built, heavy-jawed, and ultimately unsuccessful lineage named collectively as Paranthropus. According to the same author, a second shake-up occurred between 1.9 and 1.6 million years ago. An even larger-brained and more carnivorous species, *Homo erectus* (called by some scientists *Homo ergaster*; *ergaster* in Greek meaning "workman"), appeared on the scene. Its taller skeleton was nearly indistinguishable from that of modern humans. The species *Homo* was shown to be the first to migrate from Africa to populate Southeast Asia and Europe. Stone tool technology also progressed to a major upgrade: the first hand axes appeared, with large blades carefully shaped on both sides. The creatures that adapted to climate shifts—those that showed flexibility in what they ate and where they lived—were the ones that prospered. Each of the "big five" mass extinctions in the fossil record of life on earth during the past 540 million years was accompanied by environmental disruption.

During each of these periods, 50 to 90 percent of all species perished, only to be followed by bursts of new, different species. We know now that African climate history has been in continuous swings between wetter and drier times.[58]

Peter Ungar described the research of paleontologists Mark Muslin and Martin Trauth from the University of Arkansas, providing fresh insights on how climate change shaped human evolution. They suggested that climate swings filled and emptied the spreading lakes in eastern Africa, disrupting life in the rift basins. This flux may have led to fragmentation and dispersal of hominin populations. The ability to tolerate a more variable diet would have aided survival in such turbulent times. Food prints teach us that early hominin diets varied over time and space. We most likely evolved to be flexible eaters, driven by ever-changing climates, habitats, and food availability. In other words, there was no single ancestral human diet for us to replicate. Dietary versatility allowed our ancestors to spread across the planet, capable of finding something to eat on all the earth's availability of food. These authors believed that this was the key to our evolutionary success.[59]

Curtis Marean wrote that sometime between a 195,000 and a 123,000 years ago, African *Homo sapiens* populations decreased drastically due to the cold, dry climate conditions that left much of the land uninhabitable. The author explained his hypothesis that all of us alive now descended from the population of a

---

[58] Peter B. de Menocal. "Climate and Human Evolution," Science, vol. 331, pp 540–542, 2/2011.

[59] Peter S. Ungar, "The Real Paleo Diet."_American Scientific,_July 2018.

single region which survived this catastrophe. This was the southern coast of Africa. This small area, close to the city of Cape Town, known nowadays by the scientists as Pinnacle Point 13B, probably would have been the place where humans could survive because it harbored an abundance of shellfish, mammals, and edible plants. [60]

> Here we are. We all are descending probably from a few hundred ancestors. We are looking to reach eight billion soon because of our ultra-adaptability to survive harsh climate changes and always find food, securing our existence and reproduction. Let us see how we can handle the coming climate changes and our self-destructing endeavors. Before confronting our existential problem in the last chapter, let us approach the next chapter, where we will discuss fewer known facts in human brain biology and development: transcription factors, prions, brain glia, neuropeptides, and nutrients. Dear reader, it will be another challenge for you because these scientific questions are relatively new but equally important to understand how our brain became human and why we have some of our unique conditions and diseases.

[60] Curtis W. Marean, "When the Sea Saved Humanity." Scientific American, August 2010.

CHAPTER IV

The Role of Transcription Factors, Prions, Neuropeptides, and Glia in Human Brain Growth and Evolution

"Adopt the pace of nature,
her secret is patience."
—Ralph Waldo Emerson

4.1 Transcription factors

Gene regulatory networks are complex and large-scale sets of protein interactions that play a fundamental role in cellular and tissue functions. Transcription factors are the most important players of these regulatory networks. Dr. Stefano Berto describes the critical role of transcription factors on gene regulatory networks. The most influenced networks are those involved in human frontal lobe evolution and functions, emphasizing the potential relationships between the transcription factors and those cognitive skills that make humans unique. Dr. Berto investigated how transcription factors and co-expression networks evolve in the primate prefrontal cortex. In his PhD dissertation, cited below, Dr. Berto analyzed data on genome-wide expression from prefrontal cortex samples from humans, chimpanzees, and rhesus macaques. He first pinpointed the genes that are changed explicitly in each species. In total, he identified 645 genes coding for transcription factors that show lineage-specific expression, among them 134 known to be involved in brain development, functions, and diseases. Dr. Berto found that the human brains' prefrontal cortex displays the most evolutionary changes. He also discovered that one of the many transcription factors helps our hippocampus convert short-term to long-term memory.[61] Besides Dr. Berto's study, I could not find other large-scale studies in the role of transcription factors for human brain development, growth, and evolution.

4.2 Prions in brain biology

Prions are misfolded proteins that can transmit this misfolded, not normally functional state onto normal variants of the same normal protein. The name "prion" was first coined by Dr. Steven Prusiner in the early 1980s.

Historically, the diseases known today to be caused by pathological "prions" were initially suspected to be caused by "slowly infectious viruses." The fascinating story of "slow virus diseases" dates from 1956. The discovery of these "new diseases" was mainly due to the enthusiasm and dedication of another revolutionary in medicine, Dr. Daniel Gajdusek.

[61] Stefano Berto,_"Transcription Factors Networks Play a Key Role in Human Brain Evolution and Disorders," Dissertation, 2016.

The enormous efforts of Dr. Gajdusek and his Australian colleague started in 1956 in the mountains and valleys of New Guinea. They passionately investigated the peculiar spongiform vacuolar brain disease developed in New Guinea's native inhabitants. They discovered that some natives of New Guinea suffered from an unclear, progressive, and fatal disorder called by natives' *kuru* (meaning "shaking death" in the Fore language). The disease usually developed months or years after performing the traditional ritual of eating the body and brain of a dead relative. Happily, after the enthusiastic work of Dr. Gajdusek and his Australian colleague and after many further studies, this old cannibalistic tradition came to an end. No more kuru disease has since been observed in New Guinea or anywhere else on our planet. Their pioneer work is a part of numerous other articles and books for which Dr. Gajdusek received a Nobel Prize in Medicine in 1977.

In the 1970s, another scientist and neurologist from San Francisco, Dr. Stanley Prusiner, was also deeply involved in finding the elusive "agent" of another spongiform encephalopathy called Creutzfeldt–Jakob disease. In his book *Memory and Madness*, cited below, Dr. Prusiner reported about his first contact with a Creutzfeldt–Jacob disease patient when he was in his neurology residency program. His book vividly describes all the difficulties in his and his coworkers' path to the discovery of prions and their role in physiology and diseases. In his book, Dr. Prusiner recounted his scientific life, from being a medical student to 2014, including the excitement of being recognized with a Nobel Prize in Medicine and

Physiology in 1997. Most of the scientific community, including Dr. Gajdusek, reacted with intense criticism when Dr. Pursiner coined the name "prion" for the first time in 1980. This was because most scientists at that time believed that spongiform encephalopathy in animals and humans, including kuru and bovine encephalopathy, was caused by "slow viruses." The size of the viruses was believed to be 1/10 to 1/100 the size of a bacterium. The many opponents of Dr. Prusiner argued, "Why do we need another self-multiplying organism? How can proteins multiply themselves?" Dr. Prusiner and his team, joined gradually by many other teams, had given the answer to these and many more questions since then. Dr. Prusiner, in his book *Memory and Madness*, reported the history of his and his coworkers' efforts to find answers to all the above and even more questions, including the involvement of prions in forming our memories.[62]

The biggest scientific surprise was that we all have "normal prions" in practically all our cells and mostly in the brain. They are produced by a separate chromosomal gene. Dr. Prusiner and his team named this normal prion PrPc. Researchers found that patients with kuru in New Guinea and those with Alzheimer's disease have the PrP27-30 variant of the prion gene. Later, it was found that the normal PrPc protein exists isolated, without forming clusters of polymers. Instead, pathological prions like PrP27-30 have polymerase that transforms the isolated prions into cluster of prion polymers called amyloid. Enzymes called proteases destroy the normal prion PrPc but not the abnormal PrP27-30. They found that PrPc

---

[62] Stanley Pursiner, MD. *Madness and Memory. The Discovery of Prions—a New Biological Principle of Disease,* Yale University Press, 2014.

has a high alpha-helical content compared with PrP27-30, formed mainly by the non-dissolving amyloid.

> Presently, researchers concluded that most neurodegenerative disorders in humans like Alzheimer's disease, Parkinson's disease, and ALS (amyotrophic lateral sclerosis) are due to the aggregation of pathological prion-like molecules. This is currently a focus of hot research, and you, dear reader, will soon see and hear far more information about this important subject. I suspect deeply that this future research will also demonstrate that our gut microbiome is closely involved in forming pathological prions, especially in our brains. Wait, we will talk about the microbiome in the next chapter. However, the role of the normal and pathological prions in human brain growth and evolution is still unknown and remains to be investigated.

In February 2018, I had the chance to attend Dr. Steven Prusiner's lecture during the Annual Nantz National Alzheimer Center Symposium in Houston. He was kind enough to sign my copy of his book *Madness and Memory* with a cast on his right hand. Thank you, Dr. Prusiner!

One of these normally existing prion-like molecules is CREB (cAMP response element binding) protein. CREB is connected to the gene promotor and was initially identified in one type of sea slugs called Aplysia. Dr. Eric Kandel emphasized in his scientific bestseller *In Search of Memories*, cited below, that CREB is an important switch from short-term to long-term memory. It does not degrade as fast as other proteins. Genes for prions code for the normal recessive forms of prions. These normal recessive-form prions can be converted to the dominant form: self-propagating molecules in synapses as discovered in 2001 by Dr. Eric R. Kandel and his postgraduate student Kausik Si. The inactive form of CREB is transformed into an active form under the influence of the mediator serotonin. The above introduction about CREB stems from the groundbreaking book by Dr. Kandel, which was fundamental for me as a clinician and neurobiologist. This book on memory, CREB, and other molecules involved in memory acquisition and consolidation is always at hand on my bookshelves. Dr. Kandel described his work and struggle in studying memory mechanisms after he switched from being a clinical psychiatrist to studying memory in the 1960s. He was awarded a Nobel Prize because of his numerous scientific contributions, especially on studying the intimate mechanism of memories.[63]

> Dr. Kandel described a scene in his book that is vividly engraved in my memory. One early morning in 2001, Dr. Kandel's fellow and close collaborator, Dr. Si, knocked on Dr. Kandel's Columbia University office door. Extremely excited, Dr. Si shared with Dr. Kandel that he just observed how "the CREB molecule looks just like a prion." This highly significant discovery led Dr. Kandel and his team to the idea that prions could also be the basis of our memory. What a collaboration and what a revelation! Dr. Kandel and Dr. Si realized that nature uses its old best "creations" such as normal prion-like molecules to build new powers—like memory. After defending his PhD thesis under Dr. Kandel's supervision, Dr. Si founded his own laboratory, continuing to work on prion-like molecules, memories, and related diseases. To my personal knowledge, as of 2015, Dr. Kandel still works in his office at Columbia University and can be seen with his wife on their daily walks in Central Park.

Fukuchi and his team added that CREB in humans appears to be a mutation that helps stabilize the memory. When people learn something, the soluble prion-like molecules CREB turn into an aggregate prion. This aggregated CREB plays a role of a transcription factor that is necessary for the continuous synthesis of proteins to sustain our memory. If CREB is not activated, the synapse will collapse, and no memory will be formed. CREB prion-like aggregates renew themselves

[63] Eric Kandel. In Search of Memories. The Emergence of a New Science of Mind. NY: W.W. Norton, 2006.

continuously, recruiting newly made soluble prions into aggregates. This ongoing maintenance is crucial for normal functioning of memory. It was found that CREB is decreased in depression and in Alzheimer's patients. [64]

Each human is brought up with different experiences. Consequently, the architecture of each person's brain is unique. I completely agree with the opinion of physicist Michio Kaku that "the two greatest mysteries in all nature are the mind and the universe." I was excited to learn that a highly accomplished physicist as Michio Kaku was also interested in the brain and the role of emotions in forming our memories. He wrote that emotions can elevate the gene activator CREB1 and decrease the gene repressor CREB2. He established further that "more CREB activator proteins circulate in the brain after long term memories are formed." Kaku dug deeper into brain biochemistry, revealing that other neuro mediators like adrenaline are needed too for memory formation.[65]

> I sincerely believe in Michio Kaku's statement that "our mind and the cosmos are the two greatest mysteries." To illustrate Dr. Kaku's and my humble opinion, I spent my free time in the last seven years studying two relatively new hormones, neuromodulators, and trophic factors. They are the primary regulators of CREB secretion. For most medical practitioners, this next passage will be a novelty as it is for you, dear reader. Welcome to these novelties: PACAP and VIP!

[64] Fukuchi, M et al., "Gene cells," 2016, 21(8), 921–9.

[65] Michio Kaku. The Future of the Mind. The Scientific Quest to Understand, Enhance, and Empower the Mind. NY: Random House, 2014.

4.3 PACAP and VIP in human brain growth and function

PACAP (pituitary adenylate cyclase activating polypeptide) and VIP (vasoactive intestinal peptide) are two of the ten neuropeptides from the secretin hormone family. Secretin was the first hormone discovered in 1902 by Bayliss and Starling. Secretin is mainly secreted by the duodenum (the initial part of the bowel), and it has three main functions: regulation of gastric acid, regulation of pancreatic bicarbonate, and osmoregulation. It is believed that all the secretin family neuropeptides (enumerated below) evolved in vertebrates by gene or genome duplication. Currently, the ten known peptides belonging to the secretin hormone family are as follows:

- Secretin
- *VIP (vasoactive intestinal peptide)*
- *PACAP (pituitary adenylate cyclase activating polypeptide)*
- PHM (peptide histidine methionine)
- PRP (PACAP-related peptide)
- GHRH (growth hormone–releasing hormone)
- Glucagon
- GLP1 (glucagon-like peptide 1)
- GLP2 (glucagon-like peptide 2)
- GIP (glucose-dependent insulinotropic peptide)

In this chapter, I will focus on the last-discovered and less-known VIP and PACAP.

Hirabayshi and colleagues recently emphasized that all hormones from the secretin family, including VIP and PACAP, connect with the G protein–coupled receptors type 2 (GPCRs type 2). These receptors are localized on the cell membranes of virtually any cell. This hormone-receptor connection initiates the cascade of actions leading to increasing the intracellular cAMP and consequently regulating CREB. The GPCRs may have had a common origin six hundred million years ago with the appearance of primitive chordate animals. Now these receptors are found on the cell membranes on organs and in bigger quantities, mainly in the brain. PACAP and VIP functions include controlling anterior pituitary hormone secretion, vasodilatation, insulin secretion, adrenaline secretion, and immunosuppression. In the central nervous system, they also act as a neurotransmitter. PACAP, and to a lesser extent VIP, exerts a neuroprotective effect in response to cerebral brain ischemia, Parkinson's disease, traumatic brain injury, and spinal cord injury. It was discovered in 2003 that PACAP acts on the proliferation of astroglia cells too. In 2006, researchers found that PACAP also influences the neuronal differentiation of neural progenitor cells.[66]

Yin-giu Wang and colleagues confirmed that the PACAP precursor gene underwent an accelerated evolution in the human lineage since divergence from the chimpanzees. The amino acid substitution rate in humans was found to be at least seven times faster than in other mammalian species. Eleven human-specific amino acid changes were identified in the PACAP precursors. Their data suggested that the PACAP precursor gene underwent adaptive changes during human origins and may have contributed to the formation of human cognition. PACAP and VIP increase the formation of the normal membrane

---

[66] Hirabayshi et al., "Discovery of PACAP and Its Receptors in the Brain." J. Headache Pain, 2018. 19,1,28–

prions (PrPc), and PrPc increases DNA repair, improving neuronal survival. Lately, diseases such as Alzheimer's, Parkinson's, and ALS are attributed to pathological (misfolded) proteins called prion-like molecules. Glial cells are highly sensitive to PACAP. Microglia activation increases the formation of misfolded prions. The human genome encodes roughly 1,500 transcription factors, including the VIP and PACAP stimulation product—CREB. As already mentioned in the previous CREB section, activated CREB induces gene expression. CREB can bind and modulate 1,900 regions in the human genome. CREB is activated by VIP and PACAP.[67]

However, the role of VIP and PACAP in human brain evolution has not been yet evaluated. Below, I will synthesize some of the most important studies presented at the Fourteenth International VIP/PCAP Symposium in Los Angeles in November 2019, where I displayed a poster on the literature reviewing the role of VIP and PACAP in neurological disorders. I also had the chance to meet many enthusiastic researchers in our common area of interest. Here are some highlights from this symposium:

Dr. Illiana Gozes and her team discovered the astrocyte-dependent neuroprotective protein (ADNP) in 1999. ADNP is produced by astroglia cells under the stimulation of PACAP and VIP. Dr. Gozes and her team from Israel started the synthesis of the ADNP-snipped peptide NAP under the pharmacological name Davunetide, which is presently in a phase II trial for treatment of children with autism. NAP is a synthetic octapeptide (eight peptide) but can mimic 33 percent of the ADNP function.

[67] Wang, Y. et al. Genetics, 170(2), June 2005, 801–806.

ADNP is vital for the development of normal synapses and neuroplasticity. Parents of children with autism participated during the first day of the symposium and asked many questions about NAP and its treatment potential. Dr. Gozes and her team members answered to the parents that the primary results from using NAP were safe and optimistic. They also revealed to the parents that they observed an increase in the IQ in some autistic children. In their presentations, Dr. Gozes and her team reported that they found ten different mutations of the large ADNP molecule in autistic children. Interestingly, they also observed thousands of mutations in Alzheimer's disease patients. It was also reported that PACAP deficiency leads to schizophrenia; therefore, PACAP agonists are being considered for treatment of that condition.

Over the last twenty years, Professor Dora Reglodi created a strong team in Pécz University, Hungary, to investigate VIP and PACAP function. I learned that previously, Dr. Reglodi worked in the team of the discoverer of PACAP, Dr. Akira Arimura, in New Orleans, USA. PACAP was discovered in 1989. Interestingly, PACAP and VIP (discovered in 1979) have attracted the attention of very few researchers in the United States. It took enthusiastic researchers mainly from France, Hungary, Italy, Japan, and Hong Kong to start working on these two hormones and organize many international symposia on VIP/PACAP.

Professor Reglodi, Dr. Tomasz, and their team in Hungary continued to intensively study VIP and PACAP in health and disease. Many researchers on their team presented their results at the symposium. One of the researchers from Hungary presented that serum PACAP is increased in pregnant and lactating women. They also showed that in human breast milk, PACAP is present in a higher concentration compared with nonhuman primates. Dr. Tomasz reported that PACAP is increased in the serum of PTSD patients.

In the photograph below, my poster was honored by the leading researchers in the ADNP and VIP/PCAP field from Israel and Hungary. From left to right: Dr. Tomasz, Dr. Gozes, and Dr. Reglodi.

I also met and shared information with French researcher Dr. Vaudry, as well as several Italian, Japanese, and Hong Kong researchers. Surprisingly, among the one hundred researchers from other countries, I met only few American ones. I hope that my second homeland, USA, a country with an enormous scientific potential, will join these researchers to better investigate and understand VIP and PACAP and implement this knowledge for clinical purposes.

Indeed, researchers in the USA created the synthetic VIP (aviptadil), which the FDA approved twenty years ago to treat certain lung diseases. I expect aviptadil to be soon officially approved by the FDA for treatment of patients with advanced COVID-19 lung disease. VIP was discovered in 1979. Its function as a trophic factor, hormone, mediator, and mitochondrial enhancer is better known than PACAP's function. VIP function overlaps that of PACAP 80 percent. VIP operates mainly in the immune system and the lungs, while PACAP acts mainly in the central nervous system, peripheral nerve system, autonomic nerve systems, and the lungs. As I emphasized previously, both hormones have receptors in practically all organs in the human body.

In the summer of 2020, the Houston Methodist Hospital reported a trial of aviptadil. This trial was performed on eight patients with very advanced COVID-19 disease. The report was that all eight patients survived after receiving eight hours of intravenous aviptadil infusion daily for three days. These eight patients had been treated earlier, unsuccessfully, with all known methods, including ECMO (extracorporeal membrane oxygenation). Aviptadil has a long half-life compared with VIP and PACAP hormones with a half-life of less than five minutes. We just heard that the pilot placebo-controlled study of 196 patients with terminal COVID-19 disease using intravenous aviptadil infusions had been completed. We expect the results to become public knowledge at any moment.

Despite VIP and PACAP having significant differences in humans compared with nonhuman primates, there are no studies yet focusing on their importance in human brain growth and evolution. I hope this book will stimulate chemists, biochemists, and other specialists with gas chromatographs in their labs to interact and collaborate with each other and clinicians like me to conduct a more detailed investigation of VIP and PACAP in health, diseases, and human brain growth and evolution. I have been searching for such a collaboration since 2015. Until now, I have had no success. I believe that these two "super hormones and super trophic factors" are too complex to understand without full multidisciplinary collaboration.

4.4 The crucial significance of the frontal and prefrontal lobes in human evolution

Iain McGilchrist reported that the frontal lobe represents 7 percent of the canine brain, 17 percent of the lesser apes' brain, and 35 percent of the great apes' and the human brain. This author also noticed the extraordinary expansion of the human frontal lobes, the most recently evolved part of the brain. This enables us to plan, think flexibly and inventively, and, briefly, to take control of the world around us rather than to simply respond to it passively. The frontal and prefrontal lobes of our brains have made us the most powerful among animals. They also turned us, notably, into a "social animal and into an animal with a spiritual dimension." The right hemisphere appears to be deeply involved in social functioning. In most humans, speech resides in the left hemisphere. The left hemisphere yields narrow, focused attention mainly for the purpose of gathering and feeding. The right hemisphere sees the whole picture.[68]

Dr. Dean Falk found that casts of skulls of the Australopithecus (who lived 2.3 to 3 million years ago) reproduce a clear middle frontal sulcus in the dorsolateral prefrontal cortex not existing in apes. The dorsolateral surface of the middle prefrontal cortex in humans is longer. This part of the brain is associated with the executive function maintaining attention, monitoring working memory, and coordinating goal-oriented behaviors. Chimps live in groups of fifty or so, while human tribes live in groups of 150 or so. The number of the living group in different primates is known as Dunbar's number, according to paleontologist Dr. Robert Dunbar, who studied this question. The bigger number of individuals living in the same human tribe was a real advantage for increasing our social contacts.[69]

Dr. Michael Hofman reported that the surface of the human brain is about two thousand square centimeters, indicating that mammalian brains change their shape by becoming folded as they increase their size. The relative white matter volume increases with increasing brain size: from 9 percent in pygmy marmosets to about 35 percent in humans, the highest value in primates.[70] Bridget Alex synthesized in her *Discover* article that no other mammal, except humans, can adapt to any environment. She

[68] Iain McGilchrist, The Master and His Emissary. The Divided Brain and the Making of the Western World. New Haven:Yale University Press, 2009.

[69] Dean Falk, "Interpreting Sulci on Hominid Endocasts: Old Hypotheses and New Findings."_Frontiers of human Neuroscience, 01, May 2014.

[70] Michael A. Hofman, "Evolution of the human brain: when bigger is better" Front of Neuroanatomy, 8:15, 2014.

continued that the lateral parietal to frontal lobe network is very developed in humans. In other primates and mammals, the parietal lobes are primarily connected to the amygdala. In humans, however, the parietal lobes are multiconnected, increasing our focus, systematic approach, and attention, among other functions. Humans have multiple parallel parieto-frontal circuits.[71]

At the same time, this multi-connection probably creates hypersynchrony which increases, and we probably have more possibility of seizures as a "side effect" of the growth and evolution of our brain. A neurosurgeon from San Antonio, Dr. Dilup Nair, concluded his lecture that this multi-connection in the parieto-frontal lobes is a real problem for epileptologists trying to find where exactly the seizure starts. The seizures may have a frontal lobe appearance, but the seizure focus could be registered in the parietal areas.[72]

As a neurologist, I have seen many patients with different types of seizures. We usually have good or relatively good therapeutic results, treating approximately 70 percent of our seizure patients with appropriate anticonvulsant medications. The patients who failed our attempts for medical treatment are referred in general to a specialized epilepsy center. Such a center existed at the University of San Antonio for the last thirty plus years. Now among the leading neurologists at this center are professors Dr. Akos Szabo and Dr. Lola Morgan.

---

[71] Bridget Alex, "'The world is our Niche'." Discover 3/2019

[72] Nair, Dilup MD, Epileptologist, "Frontal lobe seizures". Neuroscience Grand Rounds, UTHSCSA, 3/22/2019

They and their specially trained colleagues and staff, with the substantial help of the neurosurgery department, implant subdural or deep electrodes and record the seizures on video and long-duration EEG in the university hospital setting. After a collection of sufficient amounts of seizures and a profound analysis performed by the specialized teams, the patient could be found to be a surgical candidate for treatment of the seizures. The seizures initiating from the frontal lobes are expressed by a wide variety of clinical symptoms in different patients. They are not so uniform or easily recognized as the seizures stemming from the temporal lobes.

David Eaglaman explained that every creature is wired to seek reward. The major bonus from our brain is to predict our future rewards. Social skills of people are deeply rooted in our neural circuitry. Social pain from exclusion or isolation activates the same brain regions as physical pain. We are a splendid social species, but under the microscope, it is almost impossible to tell the difference between rat neurons and human neurons.[73]

As we will see in the next section, it is not the same for the glial cells.

4.5 Glial cells and the human brain growth and evolution

"Astrocytes are the basis of humanity" - Ben Bares

[73] David Eagleman, The Brain. The Story of You. NY: Knopf Doubleday, 2015.

Astroglia (astrocytes)

Astrocytes are classes of neural cells of ectodermal origin that sustain homeostasis and provide defense for the central nervous system. They are highly diverse in form and function and demonstrate remarkable adaptable plasticity. They have many functions. In brief, they

- synthesize glycogen,
- control the blood-brain barrier,
- act as chemoreceptors, and
- together with microglia, represent the major defensive system of the central nervous system.

According to James Robertson, the progressive enlargement of the hominin brain began about two to two and a half million years ago. The changes started from the bipedal Australopithecus who had a brain comparable in size to that of the modern chimp. It is important to point out that these studies focus on the density of neurons, excluding the glia, which account for approximately 85 percent of the human neocortex. Additionally, electrical and histological studies showed no significant differences in neuronal electrical properties, neural cell types, or depth of cortical lamination among mammals. The differences were found in neurogenesis, migration, and axon guidance. All these three properties are clearly functions of the cells within the astroglia lineage. According to Robertson, the protoplasmic astrocytes, the predominant cell in mammalian gray matter, are essential for normal synaptic function and maintenance. Additionally, astrocytes are instrumental in the expression, storage, and consolidation of synaptic information from the individual synapse to global neuronal networks. Astroglia cells help build synapses. Synapses become evident within three weeks after the birth of neurons. Recent anatomical, functional, and genetic alterations in our astroglia found recently were determined to be specific to human brains. This strongly supports the critical role of astrocytes in human brain growth and evolution.[74]

According to *Science Daily*, citing an article published in *Glia*, brain astrocytes also play a starring role in long-term memory. Astrocytes generate signals of calcium and release substances known as gliotransmitter. Neurons are born following asymmetric divisions of radial glial cells. Each radial glial migrates along the extended processes of its parent cell for proper placement within the laminated neocortex The daughter cells of the radial glial cells maintain traces of their astrocytic origins even at the "peak of neurogenesis." Although humans and chimpanzees have similar gestational periods (thirty-eight weeks in humans and thirty-five weeks in chimpanzees), fetal brain volume in humans increases dramatically in the second half of gestation compared with chimpanzees. Prior to birth, the velocity of brain growth in humans is fivefold higher than in chimpanzees.

Science Daily also discusses the development of neuronal pathways. Astrocytes direct axonal trajectories by acting as guide cells of the myelin fibers to reach their correct targets. The radial glial cells subsequently were switched from neuron "production" to astroglia cells "production." This final stage is a massive production of astrocytes

[74] James Robertson. "Astrocytes and the Evolution of the Human Brain." Medical Hypotheses, 82, 2, 236–239, February 2014."

that occurs primarily after birth when the human brain enlarges the most. In addition, signals from astrocytes, along with multiple other cells, stimulate stem cells to produce new neurons. Anatomical studies showed that the volume of human protoplasmic astrocytes, the most common cell in the neocortex, is twenty-sevenfold compared with rodents. Protoplasmic astrocytes in one domain encompass ~90,000 synapses in rodents and ~2,000,000 synapses in humans. Human protoplasmic astrocytes occupy all six laminae of the cortex, where each domain is contiguous and continuous with its neighbors, resulting in three-dimensional tiling of the entire cortical gray matter. Human astrocytic processes are ten times more numerous and 2.6 times longer than those of rodents. The article in *Science Daily* continued, "Thrombospondins are astrocyte-secreted extracellular-matrix glycoproteins that control fetal synaptogenesis and neurite outgrowth. One gene of the thrombospondin family, thrombospondin 4, occurs in adult humans and was upregulated during human brain evolution."[75]

Gary Lynch also emphasized the importance of thrombospondin, a protein produced by astrocytes. According to his scientific sources, we produce many times more thrombospondin than chimps and macaques. Thrombospondin is explicitly found in the cortex. Modern human brains weigh between 1,100 and 1,500 grams (average of 1,350 grams). Our great brains and our intelligence have changed everything in the world.[76]

---

[75] *Science Daily*, 8/23, 2019; Glia, 7/26/2019.

[76] Gary Lynch and Richard Granger: BIG BRAIN. The Origins and Future of Human Intelligence. London: Palgrave, Macmillan, 2008.

Here is the moment to tell the fascinating postmortem story of the long, strange journey of Einstein's brain told by Brian Burrell. Albert Einstein died at age sixty-seven from a burst aortic aneurysm in 1955. The pathologist, Dr. Thomas Harvey, removed Einstein's brain seven hours after his death. The brain reportedly weighed 1230 grams. The pathologist kept the brain in 10 percent formalin. Dr. Harvey had no legal right to do this, but he managed to solicit the permission and the blessing to do it in the name of science from Einstein's son Hans Einstein. Because Dr. Harvey refused to return the brain to the hospital where the autopsy was performed, he was dismissed from that hospital. Instead, Dr. Harvey took the "precious Einstein brain" to Philadelphia where a skilled technician sectioned it into over two hundred pieces fixed in collodion. Reportedly, Dr. Harvey gave six pieces of the brain to Einstein's primary physician, Dr. Harry Zimmerman, and placed the remaining pieces in two formalin-filled jars, which he stored in the basement of his house in Princeton, New Jersey. Later, Dr. Harvey moved from state to state, giving away a few pieces to researchers here and there, but no major breakthroughs in the search for Einstein's genius were found until 2005. [77]

---

[77] Brian Burrell. *Postcards from the Brain Museum: The Improbable Search for Meaning of the Matter of Famous Minds.* NY: Broadway Books, 2006.

> In 2010, at a glia meeting in Birmingham, Alabama, I learned how this story unfolded. One piece of the right parietal lobe of Einstein's brain was given by Dr. Harvey to astroglia investigators, who found, with the help of special brain coloration, that Einstein's right parietal brain had four times more astroglia cells than the usual postmortem area of a male brain at his age. Persistence and curiosity paid off. There will probably be more discoveries in the remaining pieces of Einstein's brain which are waiting in the formalin for other curious researchers.

According to Suzana Herculano-Houzel, human brains are awesome. Our brains are seven times larger than they should be in proportion to the size of our bodies. Haller's rule of 1962 claims that larger animals have larger brains. But we became remarkable without even being special. The primate cortical-to-cerebellum neuron ratio is 1:4.2. In the 1.2-kilogram elephant brain, this ratio is 1:45. Primates show a much faster addition of numbers of neurons in the cerebral cortex than in the rest of the brain. Dr. Herculano-Houzel and her team estimated that we have sixteen billion neurons in the cortex, sixty-nine billion in the cerebellum, and one billion in the rest of the brain. We fit all these eighty-six billion neurons in an extraordinarily small space. Baboons have only eleven billion cortical neurons out of ninety-one billion neurons altogether. More than the forty-one animal species' brains tested in Dr. Herculano-Houzel's lab have more neurons than glial cells. The ratio of neurons to glial cells in humans is 1:1, the same as in other primates. The mass of the brain correlates with the number of glial cells. Two brains of the same size will have the same number of glial cells. In primates, cortical neurons stopped growing; they just increased in numbers. Human neurons are the same size as the neurons of the rat or rabbit and definitively smaller than the neurons in the brains of a non-primate mammal, the agouti, for example. The smaller size of the primate neurons is a clear advantage as compared to other non-primate brains. Primate neurons stopped enlarging but became more numerous, with far more connections than any other mammals. The primate way of putting together a cerebral cortex and cerebellum neurons was also more economical in terms of volume.[78]

According to Mathew Lieberman, data from the Allen Institute showed that genes associated with glial cells are different in humans as compared to nonhuman primates. This variation could have caused the increased plasticity in the human brain. Another observation showed that the human brain's increased plasticity could be less genetically inheritable than in chimps because our brain is less developed at birth. We have a longer period during infancy and childhood when the surroundings can continue to mold our brain.[79] We keep learning.

---

[78] Herculano-Houzel, Suzana. The Human Advantage. How Our Brains Became Remarkable. Cambridge, MA: MIT Press, 2016.

[79] Mathew Lieberman. Social. Why our Brains Are Wired to Connect. New York: Crown Publishers, 2013.

We are fortunate in San Antonio to learn from one of the leading researchers in astroglia, Dr. James Lechleiter, and his laboratory. I met Dr. Lechleiter at the Glia Symposium in Birmingham, Alabama, in 2010 and continue to follow his achievements in the Medical Center of San Antonio. He founded his company, Astrocyte Pharmaceuticals, where he and his team are creating a new astrocyte-based medication for traumatic brain injury, which will probably to be used in other neurological disorders. Below is the short summary written by staff writer Jesse Pound on July 17, 2017, about the near future of this revolutionary new medication and its mechanism of action:

"Astrocyte Pharmaceuticals, the company Lechleiter founded with William Korinek of Boston, completed Monday its first round of financing that raised roughly $2.4 million in fresh capital. The fledgling company had $670,000 in seed funding from the federal government since 2014. Lechleiter hopes the drug, which is at least seven years away from market, could be easily administered by any football coach or paramedic to mitigate the effects of concussions or strokes. Astrocyte works by stimulating energy production in brain cells. It is named after the support cells in the brain that essentially hold the brain together like glue and maintain the ion balance for neurons. That balance goes haywire after a traumatic event like a stroke or a concussion, and the damaged cells release excess glutamate, causing neurons to rapidly fire. The excess glutamate and ion imbalance create a toxic environment that causes the neurons to swell up and die. The drug works by stimulating calcium production in the astrocytes, which helps the cells produce more energy and reduce glutamate. The turbo-boosted astrocytes can then do a better job of helping the neurons, stabilizing the scrambled ion balance, and lowering glutamate levels."

According to one of my favorite authors and "literature teachers," Dr. Fields, astrocytes have receptors on their surfaces for any hormone. Astrocytes also populate the hippocampus. In the hippocampus, astrocytes release several types of neuroactive substances, called gliotransmitters, that stimulate neurotransmitter receptors on neuronal synapses. In this way, astroglia directly regulate synaptic transmission. Dr. Fields wrote that "humans cheat the evolution process" by evolving our brains after we are born. The plasticity of our brain that continues into adulthood is why human evolution has exploded so far beyond that of any other organism. Myelination is also involved in the environmental molding of our brains.[80] The next section will take us to the second type of glial cells, the myelin-producing cells, called also white brain matter.

Oligodendrocytes, the myelin-producing cells

As already explained above, Herculano-Houzel and her team found that our neocortex contains sixteen billion of the total eighty-five billion neurons in the brain. For comparison, the human cerebellum, a key collection of gray matter that regulates motor control, packs about sixty-nine billion neurons into only 10 percent of brain

[80] R. Douglas Fields. The Other Brain. From Dementia to Schizophrenia, How the Discoveries about the Brain Are Revolutionizing Medicine. NY: Simon & Schuster, 2009.

mass. White matter formed by myelin-producing cells is vastly presented in human brains and very important for brain function. The oligodendrocytes communicate with neurons through multiple neuro mediators. Via signaling, oligodendrocytes have also been found to play a critical role in the learning process by producing new myelin. The myelin, covering the neuro fibers, is vital for movement, learning, and cognition. It is shown that exercise increases the oligodendrocyte stem cells, producing oligodendrocytes.

Miguel Nicoletis reported that the maturation of human white matter requires twenty years. The maturation of neurons and synapses continues for thirty to forty years. It is shown that in humans, the dorsal frontoparietal myelin myelinated fibers are developed far better than in chimps, although chimps have well-developed ventral frontoparietal pathways.[81] As mentioned before, the relative white matter volume increases with brain size, from 9 percent in pygmy marmosets to about 35 percent in humans, the highest value among primates.[82]

We, neurologists, are well familiar with white matter formed by myelin and one of the relatively frequent neurological diseases called multiple sclerosis. It impacts the oligodendrocytes producing the myelin and the way myelin stays healthy through life. Below is a short overview of this disease and other diseases that can mimic multiple sclerosis.

---

[81] Miguel Nicolelis. The True Creator of Everything. How the Human Brain Shaped the Universe as We Know It. New Haven: Yale. University Press. 2020.

[82] Michael A. Hofman: "Evolution of the human brain: when bigger is better." Front of Neuroanatomy, 2014, 8:15.

In each of our patients, before we diagnose multiple sclerosis with certainty, we need to rule out other disorders that can mimic it, such as Lyme disease, sarcoidosis, and neurological manifestations of lupus being the most common. I have experience treating MS patients since 1974 when in Bulgaria, and all over the world, we had almost no effective treatment for MS. We tried to help our patients with corticosteroids, physical therapy, and occasionally cryotherapy. Cryotherapy, a treatment with ice-cold water, was introduced and popularized mainly in Europe by Belgian professor Dr. Richard Gonsette. He and another renowned MS specialist, Professor Hans Bauer from Germany, came and consulted many of our Bulgarian patients with MS during the years 1981–1989. We have used their experience for better diagnosis and treatment of our patients.

By 1992, when I came to the USA and during my neurology residency, the development of new treatments for MS patients raised the hope for the patients and our success rate of treatments. In the last five years, there has been another boom of immunomodulatory medications with outstanding results, especially in the hands of multiple sclerosis specialists like Dr. Ann Bass, Dr. Rebecca Romero, Dr. Robin Brey in San Antonio, and Dr. Edward Fox in Austin to mention some of the local ones only. As a general neurologist, I have been fortunate to have their specialized help for some of my patients with MS.

Microglia

Microglial cells are the third type of glial cells. According to John Lieff, microglial cells account for approximately 12 percent of all cells found in the brain. They are a type of primitive yolk-sac-derived macrophages that populate the brain and the spinal cord early in the embryonal life. Macrophage cells act as the first form of active immune defense in the central nervous system. These are the "immune cells" of the central nervous system. In the fetus, microglial cells help establish neural networks. In adults, they signal for more connections. When damage of the brain occurs from infection or trauma, microglial cells are the first responders. Under normal situations, microglial cells transmit growth factors to boost neuronal growth. Healthy neurons send signals to microglial cells not to "eat" their synapses. Microglial cells are critical for the process of learning because they help rewire neuronal connections. They also signal the production of more myelin, which is vital for all learning, especially that involving physical movements. Unnecessary synapses are appropriately "eaten" (pruned) by the microglial cells.[83]

---

[83] John Lieff, The Secret language of Cells. What Biological Conversations Tell Us About the Brain-Body Connection, the Future of Medicine, and Life Itself. Dallas: Ben Bella Books, 2020.

Let me try to clarify the microglial cells' function in simpler terms. Their role is to "eat" (prune) and eliminate the old synapses marked by astroglia cells with complement Cq1. It can be called "mark of death," which indicates that this synapse should be treated as a "foreign body" and eliminated by microglial cells. This process obtained the scientific name of "pruning" from the millennial practice of pruning vineyards. All grape and wine producers have known that since Roman times and before that vineyards need to be trimmed (pruned) appropriately. In the Roman Empire, a festival marking the time of pruning was known as the Lupercalia festival. Many people in Europe and California can confirm that this tradition still exists.

As a young physician, I was fortunate to participate in one of these festivals known in Bulgaria as "Triphon Zarezan." This festival existed in old Thracia and Dacia even before Roman times. It still takes place in all the grape-producing villages in Bulgaria every year, on the first of February. In the villages around the cities of Gabrovo and Sevlievo, this festival became an apotheosis of socializing, wine tasting, and pruning the old vine branches. These cutoff branches were crafted into crowns for our heads, as probably the Romans did during the Lupercalia festival two thousand years ago. Seeing and feeling all the joy of the people around in full-bloom socialism (1975–1976) in Bulgarian villages is still very much alive in my memory. These will probably be the last synapses that my microglial cells will prune. I never knew back then that forty-five years later, I would study and work in San Antonio and write in this book about the same term, "pruning."

Interaction of glia and neurons

All three major types of glial cells mentioned above interact at every moment with the eighty-six billion human neurons to form the most complex organ ever seen. Let's see if I can express in simple language how the neurons and glia function—together.

> The neurons, astrocytes, oligodendrocytes, and microglial cells act in full synchrony as a team of four well-trained surgical teams. The neurons always need to be "clean" and their products of metabolism, mainly glutamate and potassium, need to be absorbed by the astrocytes. If not, neurons cannot function. The astrocytes absorb the glutamate produced by the neurons and transform it into glutamine. The glutamine is absorbed back by the neurons as "supplemental food." The astrocytes are the main "cleaners" and "fuel providers" for the neurons. Their activity also controls the distribution of brain circulation. More active astrocytes require more blood. Astrocytes are also a big "secretory factory." They secrete brain cholesterol and many substances vital for the brain called gliotransmitters. They also secrete the complement C1q that attaches to old and nonfunctional synapses, making them "interpreted" by the vigilant microglial cells as a "foreign invader in our brain." The resting microglial and patrolling microglial cells become highly voracious. They engulf and destroy the marked with Cq1 synapses. This phenomenon is happening at any moment in our brains. As explained earlier, this process is called pruning. The microglial cells clean away the old synapses and enable the neurons to grow new ones.

Clinical application of neuron-glia interaction. Now let us review autism, schizophrenia, and Alzheimer's disease from the point of view of the interactions of glia and neurons.

We can explain *autism* as insufficient or ineffective synaptic pruning. Pruning in humans is shown to be maximally expressed at age ten to twelve when approximately half of the synapses need to be destroyed, "pruned." In patients with autism, this may signify that astrocytes secrete less complement C1q and fewer synapses will be marked to be destroyed by the microglial cells. Thus, the autistic brain will be "bushy," with far more unnecessary synapses that will interfere with the normal functions of the brain. These children are easily agitated and upset by loud music or speech, and they overreact and exhibit intrusive behavior and many more different symptoms, according to the level and location of the defect. It is believed that autistic children and children diagnosed with the autistic spectrum disorder, including the milder form of autism called Asperger's syndrome, have suffered defective pruning. Despite the enormous literature published regarding this problem, still, nobody pinpoints the root cause for this abnormal development in the brain. By recent combined statistics, one out of forty-three children in the United States is diagnosed presently with autistic spectrum disorder.

In *schizophrenia*, the process is probably opposite to that described above about autism. To simplify, I will name this opposite process "over-pruning." Usually, the process of "over-pruning" is clinically expressed in the early twenties and often provoked by stress or infectious factors.

Over-pruning kills many functional synapses, and depending on the level and locus, we can observe a variety of symptoms in different patients with schizophrenia. Indeed, experts in schizophrenia have found hundreds of mutations, making the discussion of the causes of schizophrenia extremely complex and out of the scope of this book.

In *Alzheimer's disease*, the answers are even more complicated because there are many known and even more unknown factors. Thinking in the same train of glia-neuron interaction, we would expect in Alzheimer's disease a slow, progressive "over-pruning" of synapses, especially in the hippocampus, where we form recent memories. Usually, the recent memories need a good, deep night's sleep to be "uplifted" to the frontal cortex and incorporated as permanent or almost-permanent memories. At this point, I need to emphasize explicitly, dear reader, that the above description is an oversimplification of exceptionally complicated medical problems.

I am merely trying to highlight the brain also as a complex, immune, and endocrine organ, not only focusing on the already well-known role of neurons and myelin. It will not be easy for further researchers. Presently, we found ourselves in mountains of experiments and information that need to be classified and studied. I am again lucky to be working and living in San Antonio. We have the Glenn Biggs Alzheimer's Institute in San Antonio under the directorship of Professor Seshadri and with the help of the neurology department of UTHSCSA.

4.6 Apolipoproteins (APOE4, APOE3, and APOE2) and their role in human brain growth and development

According to Dr. Dale Bredesen, apolipoproteins in the brain are produced mainly by astrocytes, and some are produced from microglial cells. The macrophages in the liver produce the apolipoproteins intended to serve the rest of our body. It is estimated that the APOE4 gene appeared in primates seven million years ago, and until 220,000 years ago, all our ancestors carried exclusively APOE4. Consequently, a new allele of APOE4 first appeared, now called the APOE3 gene. Another allele, now called the APOE2 gene, appeared about eighty thousand years ago. Dr. Bredesen emphasized that carriers of the gene APOE4 have stronger antibacterial and antiparasitic immune responses, but they live shorter lives by two to three years. Twenty-five percent of the present population has one or two APOE4 genes, and according to scientific findings, the carriers of APOE4 have several times higher risk of developing Alzheimer's disease. It is found that all other primates have preserved the APOE4 exclusive status, which makes us believe that the APOE3 and APOE2 genes (mutated only in one nucleotide as compared to APOE4) had probably evolved because humans started to use fire and started cooking their food. This would decrease the chance of being exposed to infection and probably would lead to a gradual substitution of the APOE4 gene by the formation of the new alleles APOE3 and APOE2. Dr. Bredesen also believes that these two uniquely human alleles would provide us with a longer life span and less chance of Alzheimer's disease.[84]

4.7 Nutrients: Focus on the role of omega-3 fatty acids in human brain growth

In *Psychology Today* 2019, the author reviewed the role of omega-3 fatty acid in brain health. The review article also emphasized that the brains of most mammals reach full size within months or a few years, but human brains continue to grow for the first twenty to twenty-five years of age. The growth of the brain in the embryo and early in infancy is mainly accomplished by incorporation of a group of polyunsaturated fatty acids known as omega-3 fatty acids. There are three major omega-3 fatty acids: ALA (alpha-linolenic acid), EPA (eicosapentaenoic acid), and DHA (docosahexaenoic acid). ALA, EPA, and DHA are essential nutrients for us because human bodies cannot synthesize them. Therefore, we need to receive them from our nutrition. We find ALA in plants, including walnuts, flaxseed, and canola oil. EPA and DHA come exclusively from marine sources, including sardines and algae. DHA is the most abundant omega-3 fatty acid and is a key player in brain development.[85]

Robert McNamara and colleagues found out that DHA balances the inflammatory effect of arachidonic acid and omega-6 fatty acids. They also established that brain regions with high synaptic density have higher DHA concentration. One such area is the frontal cortex,

---

[84] Dale Bredesen. The End of Alzheimer's. Program. The first Protocol to Enhance Cognition and Reverse Decline at Any Age. London: Penguin, 2020.

[85] *Psychology Today*, 2019.

the center of planning and decision-making and other complex cognitive functions. Fish and other "waterside" food are rich in DHA.[86]

Stephen Cunanne and Kathlyn Steward first launched the "waterside hypothesis." Their book explains that people first migrated only around the coast. Humans found there "shore-based" food rich in DHA as well as other essential nutrients such as iron, zinc, magnesium, selenium, and copper. They also found that human babies' fat is three to four times richer in DHA than that of chimp's babies. Human breast milk is also very rich in DHA.[87]

Ruth Kassinger emphasized that our brains need a lot of nutrients, such as iodine. Most iodine comes from volcanoes. The iodization of water started in the early twentieth century. Before, the occurrence of cretinism and goiter was frequent. The daily requirement of iodine is 150 micrograms. Lactating women need two times more. The author described in detail that DHA is a constituent of nerve cell membranes, concentrated in synapse areas, and that DHA triggers more than one hundred genes vital for brain growth. Ancient humans living alone on shores had to consume more DHA from marine food.[88] Below is Bret Stetka's summary of the opinion of Dr. Curtis Marean, expressed in several sources, regarding the importance of consuming food rich in DHA for the growth of our brains:

---

86 *American Journal of Clinical Nutrition*, v.91, 4, April 2010, 1060–1067.

87 *Human Brain Evolution. The Influence of Fresh Water and Marine Food Resources.* Edited by Stephen C. Cunnane and Kathlyn M. Stewart. Hoboken, NJ: Wiley Blackwell, 2010.

88 Ruth Kassinger. SLIME. How Algae Created Us, Plague Us and Just Might Save Us. Boston: Mariner Books. 2019

Approximately 165,000 years ago, extremely cold temperatures and drought established themselves over almost all of the earth, except for a few coastal areas in South Africa. According to Dr. Curtis Marean, everybody from our species perished, except some several hundred people, including a few hundred women of childbearing age. They found refuge in the caverns of Cape Agulhas, a rocky headland in Western Cape (Pinnacle Point). Dr. Marean and his team spent twenty years excavating in that cape in South Africa. According to him, the Indian Ocean had a hot current carrying algae, mollusks, fish, and other marine food rich in DHA, helping these few hundreds of humans to survive and thrive. This fortunate moment in our history is believed to have helped the humans in this area of South Africa grow more complex brains, while all other people in Africa suffered from famine and ultra-dry conditions. Dr. Marean believed that these few hundred women of childbearing age in that area are our common ancestors.[89]

[89] Bret Stetka, *American Science*, March 1, 2016, describing the Curtis Marean 20 years of research.

CHAPTER V

The Role of the Microbiome in Human Brain Growth and Evolution

"Learning never exhausts the mind."
—Leonardo da Vinci

Our microbiome has one hundred times more genes than we have

My initial medical education in Bulgaria, a country between the West and the East, helped me become familiar with both Western and Eastern medicines. Even today, my medical curiosity for new information is very much alive and motivates me and colleagues from San Antonio to form a medical group, focusing on the role of the microbiome in health and disease for many years now.

Our group is stimulated mainly by the early interest in this problem by our good colleague and friend, a retired psychiatrist, Dr. Vroni Hetherly. Our microbiome and medicine interest group walked regularly on Saturdays in the parks of San Antonio, discussing the still unknown role of our microbiome in health and disease. The picture below reveals some of the group members: Dr. Hetherly (left middle), Dr. Cawley (right middle), Dr. Dimitrov (behind), and his wife, Nina Danoff (left).

Bill Bryson, the American-British author of the book *The Body*, explained that we are the product of three billion years of evolutionary tweaks. Eighty percent of the air we breathe is nitrogen. Bacteria transform the inert air nitrogen into a "more sociable form" of nitrogen accepted by our organism. Microbes provide us with 10 percent of our calories by breaking down foods that we otherwise cannot use, extracting beneficial nutrients like vitamin

B2, vitamin B12, and folic acid. Humans produce twenty digestive enzymes, quite a respectable number in the animal world, but bacteria produce ten thousand digestive enzymes. Bacteria can swap genes among themselves, and they can pick up DNA from a dead neighbor. E. coli can reproduce itself seventy-two times in twenty-four hours. The life of each E. coli bacteria is twenty minutes or so. The bacteria divide or perish after that. The microbe mass on the earth is twenty-five times greater than all the animal mass. We can say that we exist at their pleasure. They do not need us at all. We would be dead in a day without them. Right now, forty thousand microbe species can claim that we are their home:

- 900 in your nostrils
- 800 inside your cheeks
- 1300 on your gums
- 36,000 in our GI tract

Thank you, microbes! Nobody could exist without you!

In his book, cited below, Bill Bryson continued deciphering some of the viruses. The name "virus" appeared in 1900 and originated from the Latin word *toxin*. The first virus was discovered on a tobacco plant. Among the hundreds of thousands of types of viruses present on our planet, only 586 species are known to be pathogenic. They cause one-third of the deaths on our planet. In 2008, a study in Switzerland showed that flu viruses could survive on paper money bills for seventeen to eighteen days if these bills are coated with human mucus. If not, the flu viruses survive a maximum of one to two days. All viruses contain a stretch of sixty-two nucleotide letters that have been found in all living things since the dawn of creation. We know several millions of fungi species too, but only three hundred of them are known to affect us.[90]

Bill Bryson shared that lipopolysaccharides, secreted by gram negative bacteria, highly influence the gene that encodes nicotinic acetylcholine receptors in humans. The acetylcholine is the mediator of the vagal nerves and the rest of the parasympathetic nerve system. The bacteria send signals through the enteroendocrine cells that make up to 1 percent of the cells lining the gut, where the largest volume of our immune cells is. These immunocompetent areas of the intestine is called gut-associated lymphatic tissue, where much of our primary immune response takes place. The invaders and toxins trigger immune cells to be active. Although genetics play a role in the risk of developing an autoimmune condition, genes are simply a blueprint determining one-third of the risk of disease development. The nongenetic environmental and other triggering factors contribute to the remaining two-thirds of the risk. Most signals from the microbiome are healthy when we feed our microbiome appropriately. However, when our nutrition habits are not in line with the needs of our microbiome, its population can change to pro-inflammatory bacteria, and other opportunistic gut bugs can take over. This "microbial imbalance" is usually more often between the ages of thirty and forty when the vagal nerve's tone decreases significantly and anti-inflammatory signals

[90] Bill Bryson. The Body. A Guide for Occupants. Canada: Penguin Random House, 2019.

stop being sent out. Positive social interaction increases vagal nerve activity too. Another helping hint would be to sleep on the right side, which is best for vagal nerve modulation. Sleeping on the back is the worst. Cold water activates the vagal nerves and, in general, all the parasympathetic system. Cryotherapy (short submersion in ice and water) is a routine used today mainly by professional athletes. Other means to stimulate the vagal nerves and, in general, our parasympathetic system will be humming or the "om" mantra, gargling, laughing, singing, playing music, exercise, sunshine, probiotics, omega-3 products, ear acupuncture/acupressure, and coffee enemas.[91]

A 2018 *Washington Post* article synthesized that there would be no life on earth without bacteria. Humans can produce only eleven out of the twenty essential amino acids; the other nine amino acids are provided by our food. Amazingly, a lot of bacteria, archaea, fungi, and protists can synthesize all twenty essential amino acids along with other components needed for their—and our—lives. The American Gut Project collected more than fifteen thousand microbiome samples from different people; none are identical. The microbiome differs just as our fingerprints. It is estimated that our microbiome can contain forty thousand species of bacteria, five million species of fungi, and three hundred thousand species of parasites. *Bifidobacterium infantis* secreted in a mother's milk inhibits the growth of all gut pathogens. Premature infants are at increased risk for necrotizing enterocolitis. C-sections and formula feedings probably increase this risk too. A leaky gut creates a hyper-permeable blood

---

[91] Bill Bryson, loc. cit.

brain barrier. Pathogens circulating in the blood can enter more easily in the brain and increase the risk of inflammation. It was shown that these pathogens activate the brain's microglial cells.

> As we know from the previous chapter, microglial cells represent 10 to 12 percent of all brain cells. Their function is as good soldiers, to keep the brain safe from foreign invaders. Once activated, microglial cells can lead to a continuous state of neuroinflammation.

The *Washington Post* article continued, "There are no tests at this point to test blood brain barrier permeability. Butyric acid (butyrate), a product of a healthy microbiome, stabilizes the blood brain barrier. Artificial sweeteners alter our microbiome. If artificial sweeteners are injected into mice, mice develop diabetes."[92]

Suzana Herculano-Houzel emphasized the fundamental role of cooking in human brain evolution. There was no way the human brain could have emerged if not for a radical change in caloric intake. Cooked food is quickly absorbed into the bloodstream. In contrast, the same foods may yield as little as 33 percent of the energy when eaten raw.[93]

Microbes in the soil. The microbiome in our gut is sometimes called inner soil. Pesticides decrease the diversity of bacteria in the soil. Sugars and carbons in plants and trees are pushed through the roots and from there into the soil. The roots secrete sugars that feed

---

[92] *Washington Post* 12/2018:

[93] Suzana Herculano-Houzel. The Human Advantage. How Our Brains Became Remarkable. Cambridge, MA: MIT Press, 2016.

the bacteria near the roots, and the bacteria in exchange provide the roots with minerals and other nutrients absorbed and distributed subsequently to all the plant. Similar events happen with the nutrients in our gut. In both cases, the bacterial diversity regenerates our body. Glyphosate (Roundup) is toxic for the gut and toxic for the soil bacteria.[94]

Bacteria also recycle in the atmosphere. Research shows that a lot of bacteria rise high in the sky, helping to form clouds. Bacteria fall back to the earth too, and the air bacteria cycle continues on and on. "Fifty percent of the world is microbiology. If you do not know microbiology, you do not know half of yourself." This statement is attributed to Dr. Sheffer.

The gut has more neurons than the brain. The so-called enteroendocrine cells allow the gut to chat with the brain via hormones. Usually, this hormonal communication can take anywhere from minutes to hours. Now researchers have discovered that enteroendocrine cells can communicate with the brain the same way neurons do by sending electrical signals in just milliseconds. The findings pose the question of how this new mechanism comes into play in conditions such as intestinal and digestive disorders. Bacteria typically produce L-lactose, which is a good food for the body and the brain. If we harbor bacteria producing D-lactose, a leaky gut is created, and the D-lactose easily penetrates the blood-brain barrier, creating conditions expressed clinically as chronic fatigue.[95]

---

[94] *Interconnected Series.* Video episode.10.

[95] *Discovery Journal* 12/2018:

Some authors believe this is only one of the mechanisms of the so-called chronic fatigue syndrome diagnosed in more than one million people in the United States. These patients are mainly women in middle age. Believe me, they are among the most hurting patients I have seen and treated. They suffer from their pains and exhaustion as well as from the poor understanding of their disease by most medical practitioners. They are the most challenging patients. Typically, they have already visited multiple physicians before someone would start studying the multitude of their health problems and diagnose the condition. In my practice, I discovered that patients who have side effects from even small doses of gabapentin/Neurontin are frequently diagnosed with chronic fatigue syndrome. The easy "passage" of gabapentin into the brain and the side effects even from small doses of gabapentin could result from the so-called leaky blood-brain barrier. It would not be surprising if these patients would have leaky gut syndrome too. They are typically frustrated because traditional medicine cannot find what is wrong with them. Physicians and other medical practitioners often tell them that "their symptoms are in their head." I have seen many frustrated patients diagnosed with chronic fatigue syndrome. The recent progress in microbiome research could give these patients more answers about their conditions. My recommendations include changing their diet, introducing the right probiotics, improving their sleep, slowly increasing daily exercise, obtaining adequate sun exposure, and coaching adequate stress response. Furthermore, after explaining the complexity of their medical condition, many of them experienced the long-expected improvement.

The probiotics and microbiome

In 1905, Stamen Grigorov from Bulgaria, age twenty-six, was a medical student in Geneva. At that time, his wife, like every Bulgarian woman, knew how to prepare homemade yogurt. Stamen was impressed by the powerful healing properties of the yogurt. One day, he took a small sample of her homemade yogurt in his luggage as he left for Geneva. With the help of researchers in Geneva and the Louis Pasteur Institute in Paris, Stamen isolated a new Lactobacillus bacterium, which he proudly named Lactobacillus Bulgaricus. The same yogurt with these gram-positive bacteria is still on the dining tables in many Bulgarian homes. In the United States, you can find the Austin-based production of Lactobacillus Bulgaricus with Texan milk (branded Bulgarian Yogurt or White Mountain) in all HEB stores in Texas. Its taste is a bit sour because there is no sugar or sweetener as in most of the other available yogurts. In the ancient Thracian language, spoken in the territory of modern Bulgaria, *yo-gurt* means "hard milk." Presumably several thousands of years ago, farmers in the Balkans discovered that when they keep sheep or cow milk in small animal skin containers around their waists (at a body temperature), the milk hardened. Nobody knew why this happened until the year 1905 when Stamen Grigorov took a sample with him for microbiological investigation in Geneva and Paris. Thank you, Dr. Grigorov!

> I remember in 1966, I was sixteen, and my parents were working in Algeria, helping Algerian people with their dental skills. My mother carried with her a small container with yogurt containing the Lactobacillus Bulgaricus culture. She prepared our homemade yogurt every day. Every evening, she used to boil the cow milk, and after it cooled to the temperature of ninety-eight degrees Fahrenheit, she added a teaspoon of the Lactobacillus Bulgaricus culture into the pot of milk. She covered the pot for the night. I never knew at that age that during the night, the bacteria multiplied to trillions until the milk hardened and was ready to be consumed in the morning. All our family ate my mom's home yogurt every day. Then, the next evening, she took a small sample of the same day's yogurt and repeated the procedure. That continued for several months until another Bulgarian family brought a fresh bacterial culture from our homeland.

Lactobacillus Bulgaricus has another valuable, a little-known effect. It can partially degrade the lactose in the milk, minimizing its allergic impact on humans, especially in older age when our enzymes breaking down the lactose become weaker. Even now, when I accidentally see my patients in San Antonio, most of them are enjoying health improvement from eating the Bulgarian yogurt daily. Some of them even claimed that they "had become addicted to it" because of their improved GI and immune systems.

The gut microbiome in health and disease

The role of the gut microbiome is critical to the development of the immune system and consequently to health and disease. After examining the microbiome of 1,054 patients and healthy controls, Professor Raes and his team found that patients with depression have a severely decreased amount of Coprococcus and Dialister bacteria. Our microbiome composition influences the enzyme that produces 5-hydroxytryptophan (serotonin). This supports the theory that microbes can influence our mood and behavior. It was found recently that 90 percent of our total serotonin is produced in the gut. Some of it is released through the vagal nerve to the brain. Depression probably starts in the brain, which influences the microbiome's production of serotonin and forms a vicious circle that cannot be interrupted without changing our diet and lifestyle. The gut microbiome composition is perhaps one of the reasons that 40 to 50 percent of people in the United States suffer or have suffered from depression or anxiety. Cognitive behavior therapy, mindfulness, religion, and meditation can also help patients with depression and anxiety. The brain-gut and gut-brain connections are vitally important and vastly understudied or underestimated, especially in traditional medical education and practice.[96]

Nathaniel Dominy and George Perry reported that humans have many more copies of the amylase gene than any ape or other mammals. As we know, amylase, secreted with our saliva, digests starch. More digested starch will bring more food to the brain and result in bigger brains.

---

[96] Jeroen Raes, Leuven, Belgium, Science 2/2019:

Eating underground tubers and bulbs may have been especially critical for the first successful humans, known as *Homo erectus.* At that time, we had the advantage of eating roots as well as bananas and other products of nature. They could have learned to cook with fire.[97]

Adam Remick reported that L-serine containing foods are essential for producing another neuron membrane component called phosphatidylserine. L-serine was recently found also in other substances like BMAA (Beta-N-methylamino-L-alanine). BMAA is found in high concentration on the island of Guam in the Pacific Ocean. It can compete with L-serine. The intrusion of the BMAA molecule will reverse the hydrophobic sites of the neuron membranes from inside to outside. This hydrophobic lipid segment of the membrane will facilitate the "aggregation of different junk" and will increase the development of the ALS-Parkinsonism-dementia syndrome, frequently seen in Guam. The ALS occurrence in Guam is reported to be one hundred times higher than in other parts of the world. Increased BMAA leads to changes in the neuron's membrane and ultimately to an increase in protein misfolding. In recent years, the increased protein misfolding is believed to contribute significantly to the development of Alzheimer's, Parkinson's, ALS, and other neurodegenerative disorders. On the other hand, researchers found that food eaten by Okinawans, those living in an island in Japan, is four times richer in L-serine than the average. Okinawa is

[97] Nathaniel Dominy and George Perry, "World Dominion Could Have Begun in the Cheeks," Nature Genetics, 2007.

one of the seven blue zones of the Earth, known for the longevity of its citizens.[98]

Alesio Fasano and Suzie Flaherty wrote that there is a continuous exchange of microorganisms from the soil to human beings, from human beings to water, air, and so on. To understand the human microbiome and its implications for health and disease, we need to consider the earth's entire ecosystem as a continuous circle of life. People from industrialized countries seem to have 15 to 30 percent fewer microbe species as compared to the gut microbiome of people of non-industrialized countries. These authors also underlined the importance of understanding the human-microbiome symbiotic relationship. Host-microbiome interaction is a two-way crosstalk. The human genome has around twenty-two thousand genes; the gut microbes have more than three million genes. Humans coevolved continuously with our microbial partners. When we die, they die too. The innate immune response is activated primarily and is believed to be more important than the adaptive immune response. The adaptive immune response is acquired individually. The gut microbiome's role is crucial to healthy development in early life. The microbiome is shaping the maturation and function of the immune system, especially the first thousand days of our lives.

Fasano and Flaherty found that patients with autoimmune disease had excess Bacteroides bacterium instead of the "good" Bifidobacterium and Lactobacillus species producing GABA. Lactobacillus species convert sugar into nutritious lactic acid. The vaginal Bartholin's

---

[98] Adam Remick, "Amino Acid L-Serine in Preventing Neurodegenerative Diseases Associated with BMAA." Pharmacy Times, September 5, 2017.

glands provide nutrition for Lactobacillus growth and lubrication. According to Ravel, quoted by Fasano and Flaherty, pregnant women with permanent vaginal Lactobacillus have less preterm birth. Babies born by Cesarean section (C-section) have a higher risk of developing chronic inflammatory diseases, including diabetes, asthma, and celiac disease. Human milk oligosaccharides in breast milk are not used directly as an energy source by the baby. Instead, these oligosaccharides in breast milk feed the "good" bacterial species that colonize the infant's intestines and are also important for the baby's brain growth. In the same way, breast milk also helps the child's long-term development of a healthy immune response. The Bifidobacterium and Lactobacillus bacteria produce the short-chain fatty acids: acetic, propionic, butyric, and lactic acids. Butyric acid has been shown to induce the sprouting of neuronal dendrites, increasing the number of synapses, reinstating processes of learning and access to long-term memories. The gut microbiome also controls the production of neurotransmitters by stimulating the enterochromaffin cells to produce serotonin or directly elaborate norepinephrine and dopamine.[99]

[99] Alessio Fasano, Susie Flaherty. Gut Feelings. The Microbiome and Our Health. Cambridge, MA: MIT Press, 2021.

For your information, dear reader, a bacterium in our gut lives thirty to forty minutes. If appropriate food and conditions exist, the bacterium divides into two daughter cells. If not, it dies—and so on in all living species for at least the past two billion years. I emphasize on the important fact that 90 percent of our serotonin is produced by our gut and only 10 percent by our brain. This can explain why neurological and psychiatric help now and in the future needs to focus also on our daily food, microbiome, and gut health. Below you can read some scientific facts about our gut-brain interactions.

Sayer Ji described how the "bad Bacteroides" in our gut activate zonulin, a protein that opens the thigh gaps between intestinal cells and creates leaky gut syndrome. Typically, the gram-negative bacteria such as Bacteroides also secrete lipopolysaccharides that increase brain inflammation and activate resting microglia. Humans presently suffer from more than eighty different known autoimmune diseases. It is estimated that these diseases affect more than twenty million US citizens. Bacteroides decrease the maturation of Treg lymphocytes and dysregulate our immune response. However, Bacteroides fragilis improves immune function. According to Ji, the microbiome is a key to evolutionary survival. Every bite we eat contains a billion microorganisms. Our bodies are connected to earth through microbes. Ji's website, GreenMedInfo.com, has been visited fifty million times since 2007. The new term "nutrigenomics" deals with the interaction between nutrients and genes. The Paleo-deficit disorder, according to Ji, represents an evolutionary mismatch like a magic human-plant coevolution. Food is essentially an epigenetic modifier of gene expression.[100]

According to scientific writer Grace Browne, the microbiome has become one of the hottest buzzwords in science, specifically the gut microbiome. Our abundant microbial inhabitants have been theorized to influence our mind and behavior and may play a role in conditions such as Parkinson's disease, Alzheimer's, depression, and autism. We do not know yet the root cause of autism, though genetic factors are thought to be involved. However, researchers implied that the gut plays a role. Much of the evidence to support the theory has come from studies on animals; for example, when scientists put fecal samples from children with autism into mice, the animals developed autism-like behaviors. These studies were thought to suggest a causal relationship between gut bacteria and the development of autism, but rodents are a poor proxy for the complexities of autism and the human mind. Other studies have found that children with autism tend to have a different microbiome makeup compared with children outside the autism spectrum. But it has never been clear whether this divergence in gut flora is a cause or an effect.

Browne also reported that a team of researchers at the University of Queensland, Australia, looked at the stool samples of almost 250 children, of which ninety-nine were diagnosed with autism. These participants also previously provided clinical and biological data to the Australian Autism Biobank and the Queensland Twin Adolescent Brain Project. Using this data and

[100] Sayer Ji. Regenerate. Unlocking Your Body's Radical Resilience Through the New Biology. Carlsbad, CA: Hay House, 2020.

comparing it to the stool samples, the researchers found that an autism diagnosis was associated with a restricted and poor diet as autistic people tend to have a sensitivity to—and dislike of—certain foods. In addition, 70 percent of children with autism tend to suffer from gut-related issues, such as constipation, diarrhea, and stomach pains. This, in its turn, was linked to a lower microbial diversity, suggesting that autism-linked behaviors may explain the differences in microbiome makeup rather than the other way around. The researchers looked at over six hundred species of bacteria identified in the gut microbiomes of the study's subjects and found only one—Romboutsia timonensis—to be associated with an autism diagnosis; the species was significantly less abundant in the autistic participants. The two datasets allowed them to look closer at the participants' diets, and they found that those of autistic people were significantly less diverse and of lower quality. Despite the lack of concrete evidence to support efficacy, early research has encouraged clinics to offer treatments for autistic people, including interventions such as probiotics, prebiotics, and fecal microbiota transplants. Fecal transplants—in which the microbes from the poop of a healthy person are administered to the patient either anally or orally—have been shown to benefit some conditions, specifically in treating Clostridium difficile colitis. This is an often debilitating, sometimes fatal, condition which arises from the overuse of antibiotics, obliterating bacterial balance in the gut. This success led to trying the treatment with fecal implants in more and more conditions—autism included.[101]

---

[101] Grace Browne. "The Gut Microbiome's Role in Autism Gets Murkier." Wired 11/19/1921.

It is estimated that presently, the local population in the Amazon River has more than fifteen hundred types of bacteria. The bacterial diversity decreases significantly in the modern world. Now scientists can count only around one thousand bacterial types in the human microbiome in the North American population. Microbiome science is developing fast, and many specialized centers are trying to determine our complete microbiome composition and microbiome genome. More than forty thousand fecal transplants are reported annually in US clinics. However, the fecal transplants are more popular in Europe and especially in Germany. The role of the microbiome in the pathogenesis of many diseases is a subject of multiple studies all over the world.

Many patients with Alzheimer's disease display increased gram-negative Bacteroides and decreased gram-negative Firmicutes leading to gut dysbiosis. Bacteroides also activate the brain microglia through secretion of lipopolysaccharides and cytokines. Both lead to increased blood brain barrier permeability, increased neuro-inflammation, reactive gliosis, and neurodegeneration. It was found that in Alzheimer's disease patients, there is also a significant decrease of the beneficial Akkermansia bacterial species, leading to weakened intestinal barrier and leaky gut syndromes.

There are also findings published from the laboratory of Mazmarian in 2016 that fecal material from Parkinson's disease patients transplanted to mice creates Parkinson's-like symptoms in these mice compared with the mice transplanted with healthy volunteers' fecal materials. Mazmarian and his team believe that certain gut bacteria are probably responsible for the development of Parkinson's disease.

My present belief is that the "bad microbiome," mainly thought the secretion of lipopolysaccharides, increases the misfolding of the alpha-synuclein to promote the development of Parkinson's disease. A similar mechanism of misfolding of the beta-amyloid and the tau protein may play a role in the development of Alzheimer's disease and other neurodegenerative disorders. Some evidence points out that our microbiome plays a significant role in the regulation of our gene expression in health and diseases. According to my literature review, this area of medical knowledge progresses slowly. We need greater collaboration among internists, GI specialists, immunologists, integrative medicine specialists, neurologists, psychiatrists, and practically all other specialists. Our patients are waiting. The role of human microbiome for our brain growth and evolution is still open to research.

CHAPTER VI

The Role of Social Life and Gene-Culture Interactions

Each of us is part saint
and part sinner.
—Edward O. Wilson

History makes no sense without
prehistory, and prehistory makes
no sense without biology.
—Edward O. Wilson

The interplay between DNA
and environment is what
makes each person unique.
—Unknown

6.1 General remarks

There is a massive volume of literature on the role of different social factors in the human society's development. This chapter will detail many of them. We will review the social and cultural factors in the light of their interaction with the genetic factors known as the gene-culture coevolution of humans.

Jen Viegas reported the results of the study of Andre Soussa and colleagues on the evaluation of the brain tissue of various primates and humans. The authors identified key elements that make the human brain unique, including cortical circuits underlying the production of the neurotransmitter dopamine. They investigated sixteen brain regions and found that the striatum is the most distinct region, where they observed the more human-specific differences in gene expression. The striatum is involved in motor coordination, reward, and decision-making. The study looked at the transcription profiles of 247 tissue samples from the brains of six humans, five chimps, and five macaques. The authors found a rare population of interneurons that produce dopamine, and this population of neurons is enriched in the human striatum. They did not report similar findings in chimp, bonobo, or gorilla brains. Surprisingly, the dopamine-producing interneurons were present in macaques and several other primates. Since dopamine plays many roles in the central nervous system, tied to cognition and behavior, humans would seem to have won the evolutionary brain jackpot. Our working memory, reflective exploratory behavior, and other cognitive skills are uniquely enhanced versus these abilities in other animals. Viegas emphasized that for chimpanzees, competition is the norm, not cooperation. Chimpanzee and other nonhuman ape mothers do not allow anyone else to interact with their newborns for about a period of six months. By contrast, human mothers allow others to interact with a newborn right after birth. That practice appears to have begun with *Homo erectus*, an ancestor who lived about 1.8 million years ago. [102]

According to Nicolas Wade, Dunbar believes that "language evolved as social grooming."[103] Wade emphasized that language unites people in larger groups. Fully articulated language emerged probably fifty thousand years ago. The next thirty thousand is believed to be the formative period of human history. Around twenty thousand years ago, the last Glacial Maximum occurred, rendering most of Northern Europe and Siberia uninhabitable. The weather improved fifteen thousand years ago. At that time, Wade believed that the first human settlements appeared. By 11,500 BC, a new kind of human society began to shape. People started to settle in fixed locations; this process is now called the first revolution. New hierarchical forms of society appeared around ten thousand years ago, indicating the beginning of the Neolithic age.

According to Wade, sedentary life seemed to precede the development of agriculture. Sedentary settlers were not always farmers. People settled at the end of the glacial period. Furthermore, a full-scale agriculture and stock raising occurred only later, around eight to nine millennia BC. Dogs were domesticated earlier. Other domesticated animals like sheep and goats were tamed around ten thousand years ago. Evolutionary adaptation initially manifested in human social behavior. The human settlement was followed by specialization, private

[102] Jen Viegas, "Comparison of primate brains reveals why human brains are unique." Mind and Brain, 11/23/2017, From the Science article of Andre Sousa et al:

[103] Quoted by Ale Jandra. Anthropology for Dummies. ISSU, 2019.

property, surplus goods, and trade. Humans developed a different relationship between genders based on family unions instead of the separate male and female hierarchies in chimps. Sex became private and personal. Domestication, including people, begun. Humans, with their special ability of speech, developed ways to live together in large groups of unrelated individuals. Once settlement began, human societies became more extensive and complex. Food surpluses, largely unknown to hunter-gatherers before, turned out to be critically important to settled human societies and opened the way to trade. People learned to treat strangers as kin, at least in reciprocal exchanges and trade. They learned to coordinate their activities.

Wade continued to describe that the first cities started springing up in Southern Mesopotamia some six thousand years ago. It is believed that writing was first invented there around 3000 BC, opening the beginning of the recorded history. One of the major forces in human communities was religion, which may have emerged almost as early as language. Trust became an essential part of human social glue. Religion initiated a mechanism for the community to exclude those who could not be trusted. Religion continues to be necessary for society's cohesion since it appeals to something deeper than reason.[104]

> In the next pages, I will focus on some physical human attributes that I believed were important for our brain growth.

[104] Nicolas Wade. Before the Dawn. Recovering the Lost History of Our Ancestors. Westminster: Penguin Books, 2006.

6.2 The role of our fat tissue in human brain growth and evolution

According to Daniel Lieberman, humans, like other mammals, mainly store surplus energy as fat. Humans are unusually "fat" compared with other mammals. Our babies are really "obese" when compared with other nonhuman primate infants. A monkey infant has 3 percent fat, but healthy human infants are born with 15 percent fat. The last trimester of human pregnancy is primarily devoted to fattening the fetus. During these three months, the fetal human brain triples in mass, but fetal fat stores increase one hundred times. Human fat further increases 25 percent during childhood. Without lots of fat, the human brain cannot grow so large, women cannot provide enough baby milk, and we will be of less endurance. The brain size did not suddenly shoot up in the human genus but increased steadily over more than one million years after the appearance of the *Homo erectus*. Since human babies have three times more fat tissue than nonhuman primates, the excess fat probably helped brain growth, especially in times of food shortage. The shortage of food was a widespread phenomenon in the lives of almost all species, especially humans. "This extra fat gave us an advantage of survival during the savagely ice ages."

Dr. Lieberman emphasized that, on the other hand, an excess of fat tissue can impact human health negatively. For example, the extra fat causes increased secretion of estrogen. Increased estrogen could be one of the major causes of the significantly high incidence of endometriosis in women than in nonhuman primates. Nowadays, statistics show that approximately 10 percent of women have or have had endometriosis. There are highly sporadic incidences of endometriosis in chimps.[105]

According to C. W. Headley, the white adipose tissue (often referred to as white fat) has two variations: visceral fat and subcutaneous fat. Visceral fat refers to the fat stored within the abdominal cavity and located near several vital organs, including the liver, intestines, and stomach. In contrast, subcutaneous fat is the fat located just beneath the skin. White fat serves its purpose, but an excess of it is often a reliable predictor for several serious illnesses. Too much subcutaneous fat will shape traditionally undesirable features, but an increased buildup of visceral fat is the one to worry about. It is associated with increased risk for conditions like diabetes, cancer, heart disease, and even depression. Conversely, brown fat, primarily found in the upper neck and upper back, helps us burn calories, generate heat, and regulate our body temperature. Lean people tend to have a much larger brown fat than overweight people. Lean people have more of the hormone adiponectin. On the contrary, obese people have leptin hormone insufficiency or more often leptin resistance at the hypothalamic level. This makes 60 percent of the US population overeat before they sense satiety. The leptin hormone is secreted by our two billion fat cells and is also known as the hormone of satiety.[106]

---

[105] Daniel E. Lieberman: The Story of the Human Body. Evolution, Health and Disease. London: Vintage Books, 2013.

[106] C. W. Headley, "The evolution of the human brain explains how we got so fat" YouTube, July, 2019

6.3 The role of physical activity in human brain growth and evolution

Herman Pontzer reported that humans have evolved to require far higher levels of exercise to be healthy. Our physiology adapted to the intensive physical activity required by hunting and gathering. Our closest relatives, the great apes, had habitually low levels of physical activity yet suffered no ill health effects from being "lazy." New research reveals that as human anatomy and behavior shifted over the past two million years, so did our physiology. We must move to survive. Movement and exercise cause the release of neurotrophic molecules, such as the brain-derived neurotrophic factor (BDNF) which promotes neurogenesis and brain growth. This neurotrophic factor also improves our memory and staves off age-related cognitive decline. Our metabolic engines have evolved to accommodate increased activity as well. Humans' maximum sustained power output is at least four times greater than that of chimpanzees. This increase stems mainly from changes in our leg muscles, which are 50 percent bigger. Our muscles have a greater proportion of "slow-twitch" fatigue-resistant fibers than the leg muscles of other nonhuman apes. We also have more red blood cells to carry oxygen to working muscles. Pontzer emphasized in his book, cited below, that we have a faster metabolism that supports our bigger brains. Exercises are not optional for humans; they are essential. Endurance exercise decreases chronic inflammation and decreases cortisol levels. At the same time, movement and exercise increase the effectiveness of our immune system and produce enzymes that help clean fat from circulating blood.[107]

Daniel Lieberman and Elizabeth Cooney recently explored "the active grandparent hypothesis." According to them, it seems we have evolved to live longer if we move more. Compared with humans, nonhuman apes are very inactive. People today are more active than an average wild chimpanzee, and it benefits our health and longevity. There might be something different about humans, and we cannot pin it yet. Moreover, does that "difference" help us become grandparents? We are living after we stop reproducing to increase our reproductive success by helping our children and grandchildren. What is the secret of human longevity? Many more anthropologists have written about this. Nevertheless, Lieberman and Cooney added physical activity to the list of reasons why humans evolved to be grandparents. One of the essential arguments in favor of physical activity is that it lowers systemic inflammation. It turns out that the main organ that regulates inflammation is our muscles, comprising 60 percent of our body. [108]

> Recent studies have shown that movement and exercise releases hundreds of signaling molecules into the body, and we are only beginning to learn the full extent of their physiological reach. Endurance exercise reduces chronic inflammation, a serious risk factor for cardiovascular diseases.

[107] Herman Pontzer, "Humans Evolved to Exercise," American Scientific, February 1, 2019.

[108] Daniel Lieberman and Elizabeth Cooney, "Life Span vs. Health Span: How physical activity wards off age-related disease. SPAN, December 9, 2021

It is well known that, before the development of agriculture and domestication of animals, humans were extreme endurance runners. One of the few ways to add meat to their daily menu was to run after an animal until that animal was exhausted. We had also been scavengers, and we supplemented our diet by breaking the bones of dead animals and sucking their bone marrow. I tried once to eat cooked bone marrow to experiment partly with the taste. I can honestly confess that I am not ready to repeat the experience.

For a millennia, bone marrow provided for our ancestors the necessary building material for a bigger and more complex brain. Lately, it was also revealed that regular exercise improves the effectiveness of our immune system, especially as we age. However, our ancestors never suspected that chasing a prey animal for one or two days would improve their survival and immune system.

I clearly remember one evening in 1980 in Sofia. I was working on my PhD in genetics in Duchenne/Becker muscular dystrophy. Muscle science at that time was in its infancy. That evening, I sat down in front of my old-fashioned mechanical typewriter. I laid out my idea in English on half a page that muscles, representing 40 percent of our body, do not exist just to move us from point A to point B. Nature is exceptionally sophisticated to use muscles just to move. I wrote that our muscles could be equally involved in other essential functions like supporting our immune and endocrine systems.

I made thirty or so copies of that half page and addressed each copy, along with a short letter, to world-renowned researchers in muscle physiology and pathology at that time. It took me a third of my monthly physician's salary to buy the necessary post stamps to mail the letter to researchers who are known to me in the USA, Great Britain, France, and other countries. I received no replies, but I believed that I was on to something important. I did not keep my original half-page typewriting but hope that some of the muscle scientists who had received my letter kept it. On arrival to the United States in 1992, I was entirely absorbed by clinical neurology and on rare occasion followed the progress of muscle physiology. I read recently the long-awaited news that science has discovered more than hundred different proteins secreted by the muscle; BDNF is only one of them. Science has a lot more to discover. I still believe our muscle-secreted proteins help our immune, nervous, and endocrine systems.

6.4 The role of language and other human communications in human brain evolution.

In "The Story of Our Brain," Steve Olson suspected that a primary language arose approximately two million years ago. Modern humans seem to have a unique capacity and desire to talk and learn new things. There are only two changes between the protein sequence of the gene FOXP2 in chimps and us, but we can talk, while the chimps cannot. The genetic studies showed that we have the same FOXP2 gene as Neanderthals. DNA studies also revealed that modern humans intermixed with Neanderthals repeatedly. Europeans carry on average 2 to 3 percent of Neanderthal genes. However, we were strong enough to withstand the rigors of different and challenging epochs; Neanderthals and Denisovans could not. Denisovans, as discussed in the second chapter of this book, carried their name because the bone of a child's pinky finger and one tooth were found in the Denisova cave in Russia, close to the Altai Mountains, at an altitude of 2,997 feet. The main chamber of the Denisova cave measures twenty-seven-by-thirty-four feet. The layers of soil in the Denisova cave where the bones were found date from thirty to fifty thousand years ago and are attributed to a hominin. The tooth found in the cave was bigger than expected for *Homo sapiens*, and it was like the teeth of Neanderthals. It is estimated that Denisovan DNA is found in approximately 5 percent of people in the islands of the northeast coast of Australia and indigenous people of Fiji and Papua New Guinea. [109]

Adam Rutherford wrote that more than 97 percent of the species ever existed on our planet are already gone. *Homo habilis* lived in East Africa between 2.1 and 1.6 million years ago. Their faces were flatter than Australopithecus (Latin for "Southern-ape-like creatures") living three million years ago. *Homo habilis* was probably the ancestors of *Homo erectus*, though they coexisted as well. *Homo sapiens* are identified by their usage of tools. Dr. Rutherford believes that the coevolution of minds, brains, and hands enabled us to create and use different tools. Tool-making animals make up less than 1 percent of all species. Many animals learn; only humans teach. We accumulate knowledge and pass it on. We transmit

[109] Steve Olson, Mapping Human History: Genes, Race, and Our Common Origins. Boston: Mariner Books, 2003.

information, not just via DNA down the generations, but also by culture in every direction. We log our knowledge and experience and share them. There is a gene-culture coevolution. Biology enables culture, and culture changes biology. Biology and cultural evolutions go together. Every journey of every human is built on thousands of years of accumulated knowledge. We create and teach almost all the time and learn at almost the same rate.

According to Dr. Rutherford's book, cited below, the fire is another pillar that shaped our gene-culture evolution. The impact of fire on the development of humankind is unique. Agriculture changed humankind irreversibly and set the foundations for the current era. It has been the dominant industry and technology for around ten thousand years. Permanent bases could be established and crops stored for years. Farming changed our bones and genes as well. In the past centuries, we have taken technology to levels of sophistication paralleled only to magic.[110]

Chat Cherwood emphasized the scientific fact that parts of the brain involved in language and cognition have significantly enlarged overtime. Von Economo cells, the basis of the so-called mirror neuron system that is pivotal in socio-emotional brain circuits, are larger in humans. The mirror neuron system, activated when viewing the actions of others, has complex circuits in humans. RNA that carries messages to instruct cells to make proteins is more active in the prefrontal cortex synapses in humans than in other primates. Human neurons produce more of the neurotransmitter dopamine. Dopamine is a neuromodulator involved in various cognitive functions. A link from the

[110] Adam Rutherford. The Human Book. A Brief History of Culture, Sex, War, and the Evolution of Us. NY: The Experiment, 2018.

motor cortex to the brain stem coordinates the larynx muscles, a circuit absent in chimps and macaques.[111]

According to Christine Kenneally, language appears to arise from a platform of abilities, some of which are shared with other animals. Complex language came into existence around fifty thousand years ago. Language is an evolutionary adaptation that changed everything. We have spoken in more than six thousand three hundred different languages. We, humans, are the only primates to have voluntary control of their larynx. Interestingly, the human advantage of larynx positioning can also put us at risk of choking.[112]

Surprisingly, for me, the latest statistics show that, on average, five thousand humans die from choking in the United States every year, although almost everyone knows how to perform the Heimlich maneuver for adults.

More interestingly, there are no reports of deaths from choking in nonhuman primates or other mammals. The high position of our larynx allows us to articulate speech, but it can easily provoke choking. Humans gain much by having articulated speech as our major means of communication. However, we can lose our lives from choking. What a beauty and a downside of human life.

[111] Chet Cherwood. "Are we wired differently?" American Scientific, 70–, September 2018.

[112] Christine Kenneally. "The cultural origins of language. What makes language distinctly human." American Scientific, 65–, September 2018.

6.5 Social behavior and cultures influence the development of the human brain

According to Mathew Lieberman, cooperation is one of the characteristics that make humans exceptional. Our brain is evolved to think socially. Perception and action activate the same "mirror neurons." In humans, functional MRI discovered the mirror neurons mainly in the frontal lobe cortex and a structure called the anterior insula. These neurons help us with language, culture, imitation, mind reading, and empathy. This area in the brains is bigger in humans, and this explains how we can live in better social harmony and cooperation. Social rewards enhance the pleasure of taking care of others. Human brain wiring motivates us to stay more connected with each other. Primates live in a world of "what" others are doing, and only humans live in a world of "why."[113]

Bouernfield and colleagues found out that apes and humans exhibit a distinctive area in the anterior insula called the fronto-insular cortex. This area is defined by the presence of clusters of von Economo neurons, also called mirror neurons, involved in social awareness. Von Economo described these neurons for the first time in 1925. In humans, these neurons are four times bigger than other neurons. The left anterior fronto-insular cortex is the most differentially expanded of the human cerebral cortex compared to the chimp, our oldest living relative.[114]

---

[113] Mathew D. Lieberman, *Social. Why our brains are wired to connect.* NY: Crown, 2014.

[114] A. L. Bauernfeind, A. A. de Sousa, „A volumetric comparison of the insular cortex and its subregions in primates," *Journal of Human Evolution*, 2013.

Here, probably, is the right place and moment to give a short overview on the much-discussed topic of consciousness. This multifaceted complex topic is itself the subject of hundreds of books and thousands of publications. As you may not know, the renowned Nobel Prize winner and codiscoverer of the DNA structure, Dr. Francis Crick, devoted the last twenty years of his life to the idea of finding the brain location of consciousness. He was helped scientifically by his main coworker, Dr. Christof Koch. The story goes that in July 2004, at the age of eight-eight, Dr. Crick became sick and reportedly continued working on correcting their last article on the location of consciousness in the brain. He kept on editing the article even in the ambulance, speeding him to the hospital. A few days later, Dr. Crick passed away in the hospital. Dr. Crick and Dr. Koch believed consciousness was localized in the same fronto-insular cortex and the neighboring structure, the claustrum. After the death of Dr. Crick, Dr. Koch continues to research consciousness and manages the Allen Institute for Consciousness, with headquarters in Seattle.

Daniel Lieberan and Michael Long wrote that the creative mind is the most potent force on earth. The neuro mediator dopamine gives us the power to create. Dopamine is the engine of curiosity, adventure, and eventually progress. Creativity is an excellent way to blend dopamine, serotonin, and norepinephrine. The human mind is a wandering mind. Fixing things gratifies us. The dopamine circuits are what make us humans. We plan. We imagine. Creativity is an innate part of our nature. The human frontal lobes contain the most dopamine receptors so behavior can be learned and rewarded. We can envision possible consequences of our behavior and actions. However, too much dopamine can lead to "productive misery," while too much serotonin and norepinephrine can lead to happy indolence. To live a happy life, we need to bring these mediators into balance. Lieberman and Long also warned us that the malfunction of the same dopamine circuits can lead us down a darker path, the path of addiction. The authors also underlined that excess dopamine mediation is found in patients with schizophrenia. There are now more than 108 gene mutations in different chromosome locations connected with the development of schizophrenia. Most of these gene mutations lead to super-active dopamine production and mediation. Modern psychiatry possesses a large pharmaceutical arsenal to mediate this dopamine super mediation and try to normalize it.[115]

According to Yuval Harari, in the *Sapiens,* success is an ability to compose fiction and cooperate. Neanderthals usually hunted alone or in small groups. The real difference between humans and primates is the mythical glue that binds us in large numbers of individuals, families, and groups. This glue has made us the masters of creation.

Harari believes that our cognitive revolution started approximately seventy thousand years ago and the agricultural revolution around twelve thousand years ago. Before the agricultural revolution, humans were hunters and foragers. They moved around, logging every month, every week, or sometimes every day. On the eve of the

---

[115] Daniel Z. Lieberman and Michael Long. The Molecule of More. How a Single Chemical in your Brain Drives Love, Sex, and Creativity—and Will and Determine the Fate of the Human Race, Dallas, TX: Ben Bella Books, 2018.

agricultural revolution, the total human population was estimated at five to eight million. Humans foraged not only for food but also for knowledge. Significant conflicts and wars began only after the agricultural revolution when people began to accumulate private property. Ten thousand years ago, wheat was a wild grass, one of many, confined to the Middle East. Suddenly, within just a few short millennia, wheat was growing all over the world. Wheat has become one of the most successful plants in the history of earth. People needed to settle next to their wheat fields. Humans deliberately burned vast areas of dense thickets and dense forest to create grasslands that more easily attracted hunted game and were better suited for human needs. They increased their food supply, and the population began to grow. Giving up nomadic life enabled women to have a child every year. Children were fed more on cereals and less on mother's milk. Child mortality soared. In most agricultural societies, at least one out of three children died before reaching the age of twenty. Evolutionary pressures have adapted the human brain to store immense quantities of botanical, zoological, topographical, and social information.

Harari emphasized that agriculture brought the necessity to handle large numbers. Mathematical ability was vital. The writing was invented as the maidservant of human consciousness, increasingly becoming its master. *Homo sapiens* are relatively physically weak animals whose advantage rests in their ability to cooperate in large numbers. What distinguishes us most is *something that* no other animals do: *art*. Throughout our evolutionary history, music and dance typically co-occurred. At a neural level, we now know that the hypothalamus, amygdala, motor cortex, and cerebellum are linked *to* both movement and emotion. In a large group, individuals with social skills would receive many benefits; they would know when to get help from others, whom to fight with, whom to trust, and whom to avoid. This emotional intelligence would have given them power over others. They created "artificial instincts" that enabled millions of strangers to collaborate effectively. This network of artificial instincts was called culture.[116] Around 2005, neuroscientists found that playing music can modulate levels of dopamine, the so-called feel-good hormone in the brain.[117]

In his previous book *The World in Six Songs, Daniel Levitin* emphasized David Huron's thesis in *Sweet Anticipation* that the musical brain might have helped prepare humans for survival. Ritual behavior is obviously innate and hardwired in humans. He believed that religion grew out of the desire to make sense of the world and that the prefrontal cortex is the seat of sociability. In humans, the prefrontal cortex evolved further to become the seat of music, language, science, art, and ultimately society.[118]

---

[116] Yuval Noah Harari, Sapiens, A Brief History of Humankind. NY: Harper Collins, 2015

[117] The ancient Greeks knew this and used music accordingly, to calm hostilities or arouse warlike passions.

[118] Daniel J. Levitin, The world in Six Songs. NY: Penguin Group. 2008.

6.6 The role of spirituality and religion in the evolution of human societies

For millennia organized religions have been the source of much of the best in creative arts.
—Edward O. Wilson

According to Jo Marchant, shamanism was the world's first religion. Only one man was painted in the Lascaux cave; everything else was animals. There were no boundaries between living and nonliving, humans and nature, earth and stars. During the Paleolithic era, we were an integral part of the natural world, sharing our environment on equal terms with other species. Later, during the Neolithic revolution, people cut those ties and became farmers, controlling and exploiting the land. We have been trying to separate ourselves from nature ever since.[119]

According to Harari, religion was the third great unifier of humankind, alongside money and empires. We are two billion Christians, one and a quarter billion Muslims, and many Buddhists, Confucians, dualists, Jews, and many other religions. As mentioned above, the first religious beliefs were mainly animalistic. People venerated different animals, and at the same time, they sacrificed them to appease their gods, named with a thousand names. Human sacrifice was also common in some ancient societies. People needed explanations why we are here, who we are, when it will rain again, what the sun and the moon are, and many more questions. Harari emphasized that the name "deity" comes from

[119] Jo Marchant. The Human Cosmos. Civilization and the Stars. NY: Dutton, 2020.

the Sanskrit word *dsius* (meaning "shining in the sky"). *Dsius* was spelled Zeus in Greek. The renowned Greek philosopher Plato was the first to launch the idea of the immortal soul. Another Greek philosopher, Pythagoras, studied in Egypt, where they believed that our souls belong to the stars and we have eternal life in the sky. At that time, the moon, the sun, and the five known planets that could be seen from Earth were believed to be gods.[120]

In the next several pages, in italic, I will try, dear reader, to introduce you to my historical research on spirituality, religion, and their influence on our knowledge and society in general.[121]

During the time of Plato, in the fourth century BC, the seven days of the week were named after the seven bodies seen in the sky:

- *Sun/Sunday – Domingo (meaning "of the Lord")*
- *Moon/Luna (Roman moon goddess) –- Monday – Lunes*
- *Mars/Tiw (Germanic god of war) – Tuesday – Martes*
- *Mercury/Wodin or Odin (Germanic supreme god) – Wednesday – Miercoles*
- *Jupiter/Thor – Thursday – Jueves*
- *Venus/Frigga (wife of Odin) – Friday – Viernes*
- *Saturn – Saturday – Sábado*

Thales of Miletus, born in 625 BC, lived in Ionia, Asia Minor (present-day Turkey). He is one of the seven sages of Greece. He was a mathematician, astronomer, and pre-Socratic philosopher. He is often referred to as the father of science. Thales is recognized for breaking away from the use of mythology to

[120] Harari. Sapiens, loc. cit.

[121] The following is my brief history of science and spirituality.

explain the world. Instead, he tried to explain natural objects and phenomena by naturalistic theories and hypotheses. One of the best-known Thales hypotheses is that "the originating principle of nature and the nature of matter is water." According to the Greek historian Herodotus, Thales predicted the solar eclipse on May 28, 585 BC. Thales was the first human to predict an event by mathematics and physics, not by mythology and gods. I believe it was because Thales was centuries ahead of his time.

Humanity, in general, was and is slow to accept new ideas and concepts. Most of the people were illiterate. It is well-known that history at that time was passed from generation to generation by songs. Homer is a renowned ancient author, a blind man who sang his compositions, the Iliad and the Odyssey, and played his stringed instrument and perhaps a lute. Thirty years ago, I read that in Serbia, there was a man who could sing all the Iliad and Odyssey like Homer did. Five centuries after the battle for Troy (the city of Troy was called Iliada in the twelfth century BC), Greek scribes began introducing vowels to the Phoenician alphabet and could transcribe the songs in letters. Nowadays, the Iliad is printed in 560 pages in English. I enjoyed reading Iliad in Bulgarian at age fifteen as part of our school lessons. I was simply mesmerized.

Thales paved the way for three well-known Greek philosophers. Their names are renowned in the history of philosophy and science: Socrates, Plato, and Aristotle. Socrates was the teacher of Plato, and Plato was the teacher of Aristotle, who was the private teacher of Alexander the Great for seven years. My eponym is easy to remember: SPA-A (Socrates Plato Aristotle–Alexander). Plato was recognized as the Greek philosopher with the most significant influence in his time and for many centuries after. He was nicknamed Plato because of his big forehead, by association with the plateau in Greek mountains.

Plato gave his lectures in the house and garden of academes; hence Plato's disciples were called the Academes sect. After learning this, I always wondered why history had attributed the glory of Plato to Academes? Why do we still say "Academy" and not "Platony"?

Despite the efforts of Thales, Socrates, Plato, Aristotle, and their disciples, Greek mythology persisted as the primary explanation of our complex world. As we know, Zeus and the twelve Greek gods had already dominated religion for several centuries before the Roman Empire became much more powerful. The Romans used the same gods but changed their names. Zeus became Jupiter, Hera became Juno, Ares became Mars, Poseidon became Neptune, Artemis became Diana, and Aphrodite became Venus. They only kept the name of the Greek god Apollo.

> *In the Balkans, the mother of Apollo was known by the name of Lato or Latona. Few people know that their Latino ancestors venerated the mother of Apollo, Lato, and they now carry her name: Latinos, Latin language, Latin America. Yes, Latin Americans, we are all connected. At that time, the world needed some order so the people would follow it. They believed in the order created by the gods. These gods controlled the skies and our spiritual and often our physical lives. Rules were imposed by the kings of the earth and reinforced by the rulers in the skies. People started living in bigger cities; in Greek,* polis *means "city," like Alexandroupolis and Philippopolis. The rulers of those cities were called police. As you can deduct, dear reader, the name "police" is still used nowadays in almost all languages, with slightly different spelling.*

One of the first monotheistic religions was Zoroastrianism. This religion was partially spread in Iran, nowadays Azerbaijan, and other neighboring countries. When I was thirty-seven, my wife and

I were invited to visit Baku, Azerbaijan's capital, by my spiritual brother, Nazim. He and his colleagues showed us the "eternal flame" close to the Caspian Sea, fed constantly by burning natural gas. Even now, the Zoroastrian religion venerates fire.

Monotheistic religions did not need Zeus or Jupiter or the eleven other gods. They needed only one "mighty God" in control of the world and our spiritual lives. Many believed the mighty God could control our physical lives too. Here are the twelve most prominent monotheistic religions:

- *Judaism*
- *Christianity*
- *Islam*
- *Hinduism*
- *Sikhism*
- *Baha'i Faith*
- *Rastafari*
- *Abrahamic religions*
- *Caodaism*
- *Mazdeism*
- *Babism*
- *Tenrikyo*

People needed answers to many questions they had, such as what the universe is, how did life on earth started, who we are, and many more. Before writing about the role of religions in building our families and societies, I would like to review the history of some monotheistic religions. I will start with the religion I am educated and know best about: Christianity.

After three centuries of continuous persecution and killing of Christians in the Roman Empire, Christianity became the official religion of the Romans. Before that, Judaism was already ten centuries old and very much alive. The Torah was read every year from beginning to end by rabbis and their followers. During these three centuries of the new millennium, the followers of Jesus Christ and his disciples built small churches. They survived mainly underground, in the catacombs of the Roman cities governed by the Roman gods. The Christian persecution continued until AD 312 when Constantine and his army won the battle of the Milvian Bridge. Constantine's mother, Elena, was one of those underground Christians.

Emperor Constantine was born in 272 in the Roman city Nasus (nowadays Nis in Serbia), only eighty miles west of the city of Sofia, where I grew up and lived until age forty-two. Constantine singled himself out as an exemplary soldier and Roman legion leader. The story goes like that. The day before the battle at the Milvian Bridge, he saw a "flaming cross" over the sun. On the very same day, Constantine ordered every soldier in his army to paint a cross on his shield. This was done just the night before the decisive battle with the other competitor for the throne of the Roman Empire, Maximilian, with his army three times as big. The next day, Constantine and his army won the battle, now remembered as the Battle of the Milvian Bridge. Seeing the cross over the sun, Constantine promised that if he won the battle, he would make Christianity the official religion of the Roman Empire. So he did.

In AD 321, Constantine chose Sunday as an official day of rest for Roman citizens. He built his 129-foot-tall statue in his new capital, Byzantium. The statue represented him naked, like the competing Olympians, with a crown of radiating rays. His face was turned toward the east. The stone statue was destroyed by an earthquake in AD 1106 AD. The inscription was "For Constantine who shines like the Sun." In AD 325, Emperor Constantine presided over the Council of Nicaea, the first major Christian assembly. The city of Nicaea is nowadays in Turkey, where the Christian authorities established the foundation of the Christian religion. In AD 364, the Roman Empire was officially split into two: the Western, with Rome as its capital, and the Eastern, with Byzantium as its capital. It is a curious fact that the signing of the official papers of this division happened in the city castle of Nasus, Constantine's birth city. Some of the ruins from the castle of Nasus (nowadays in the city of Nis, Serbia) are preserved, and everybody can enjoy this marvel of antiquity. You can see below a picture of the castle sent by my daughter Petia and her friend Milos:

After the split, both parts of the once-great Roman Empire survived very differently. Emperor Constantine realized the geographical weakness of Rome located in the plain without a real defense system. Indeed, only half a century after the split,

in AD 410, Rome and the entire Western Roman Empire was ravaged by Atilla the Hun and his army. Historians suspect that the pope managed to save Rome with diplomacy and much gold. Later, every Italian city took the defense of their cities in their own hands.

The East Roman Empire was governed directly by Emperor Constantine. He immediately saw the favorable location of the city of Byzantium on the Bosporus, situated on two continents: Europe and Asia. Therefore, he began building and reinforcing the defenses of his new eastern Roman capital city, later renamed after him, "Constantino polis." Constantinople is renamed Istanbul, the capital of Turkey, with twenty million habitants. While I was wondering in the street of Istanbul in 2013, it came to my mind that the new Turkish name I-STAN-bul came from the older Greek Con-STAN-tin-ople, meaning on both languages the "city of Constantine." The wall around Constantinople was started by Constantine, continued by descendant emperors, and finished by Emperor Justinian and Empress Theodora around AD 585. This thirty-foot-tall wall, whose ruins still exist, along with the wise politics of the Byzantine, prolonged the life of the Eastern Roman Empire for twelve hundred years. The city of Constantinople was finally taken by the Ottoman Sultan Mehmed II and his army in 1453. The sultan was twenty-two years of age.

If you read the amazing book of the Austrian writer Stefan Zweig, you will forever remember the battle of Constantinople in 1453. This book was written in German, published in English in 1940 under the title The Tide of Fortune: Twelve Historical Miniatures *and translated into many languages, including Bulgarian. One of the twelve historical "miniatures" described that famous battle in detail. The three thousand Byzantine defenders were taken by the very wise military strategy of Sultan*

Mehmed II and his army. The sultan ordered building a wooden complex structure on the hill across the Bosporus River next to the city. I had the pleasure of seeing one illustration of this battle on the stone fence on the side street close to Hagia Sophia in 2013 during our family visit to Istanbul. Thank you to this unknown artist and to the gentle Turkish man who led me to this small Istanbul side street to show me the painting. Hope this art can be still enjoyed by curious tourists.

The young sultan Mehmed II successfully planned to move three hundred ships over the hill using winches on the wooden structure. These ships bombarded the city from the unprotected bay. As a result, the young sultan achieved something that no one could do for twelve centuries: seize Constantinople.

In the seventh century AD, the world witnessed the birth and rapid development of the third major monotheistic religion. Around AD 622, the merchant and future prophet Mohamed was inspired to seek new beliefs and rules of life. I wish I could read and know more about Islam before writing about it. In this book, however, my intent is to focus on the role of the Islam religion in the preservation and spread of the knowledge of the Roman Empire through parts of Europe and Asia. Islam scholars also contributed to science at that time: Arabic scientists invented Arabic numerals and the concept of zero as a number. Islamic scholars gave birth to Algebra (from Arabic al-jabr*). They contributed to philosophy and medicine. My special thanks go to Abu Ibn Sina, also known as Avicenna, who, at the age of nineteen, was living in the cities of Bukhara and Samarkand, Persia, nowadays Uzbekistan. Avicenna was born in AD 980 and developed prominent poetry, philosophy, and medicine schools. His most important book,* The Canon of Medicine, *was the only medical book translated into Latin and studied for the next five centuries in all medical schools throughout Europe.*[122]

[122] The original manuscript of "The Canon of Medicine" is preserved in a museum in Damascus. The picture of the two pages of the book, shown above, was taken by a San Antonio pharmacist who visited the museum.

Above is a difficult-to-read copy of two pages of the original Avicenna book The Cannon of Medicine *kept in a museum in Damascus. This picture was shared with me by a good colleague and pharmacist in San Antonio who had the chance to see the original copy in the museum of Damascus. The* Canon of Medicine *replaced the older medical book of Galen written in the second century AD. Aside from developing medicine, algebra, and other sciences, followers of Islam, also called Muslims or Moors, developed strong armies and a culture that they spread throughout some parts of Asia and Europe.*

This brief review of the births of the three most popular monotheistic religions has the limited goal of emphasizing their role in consolidating the nuclear family. This consolidation undoubtedly played a role in increased security for growing children and improved communications between different individuals, families, and groups of people. Such social factors certainly influenced our brain functions and survival. I hope with all my heart that we will slowly find a common language and better understanding between all people and all religions. I belief this language will be based on a strong, solid nuclear family, mutual respect, knowledge, and humanity.

6.7 The role of sexual behavior and longevity in human brain development

Jared Diamond described that gibbon couples stay together for life, but they have sex only once a year, only for procreation. They live in rigid fidelity. On the contrary, chimps are very promiscuous. They have larger testes and a two to three times smaller penis than today's humans. Neanderthals had larger shoulders and shorter limbs, averaged five foot four, and weighed around twenty pounds more than hominids. They lived less than forty-five years. We have little information about their family life and sexual behavior. Humans have mixed strategies. Without concealed conception, the individuals in the human tribe could not collaborate. Bonds between a particular man and woman lie at the foundation of the human family. Women remain continually sexually active to reward men, bind a mate to them permanently, and take care of their babies. Humans can have many children from multiple partners. Diamond also reported that one emperor of Morocco in the nineteenth century had 888 offspring. Humans' sexual behavior and reproductive capability are a huge clue for our survival, and we are becoming the dominant species on earth. We are more adaptable, resilient, goal focused, and sexually driven than the other 2.2 million species as mapped recently by Google on onezoom.org.[123]

We also live longer.

Daniel Levitin emphasized that we do live longer. In the past one million years or so, our longevity has given us an advantage over nonhuman primates who lived and still have a life span of thirties or forties. We started boiling our water and using fire for cooking regularly probably one million years ago. We fought infections better and lived better socially. For many millennia, we had the advantage of larger families, involvement of the men in family life, and group sharing of food and information. We now live into our eighties and nineties and, occasionally, into our hundreds. This enables us to help our families and friends until our old age and to contribute a longer time to society. Abstract thinking

---

[123] Jared Diamond. The Third Chimpanzee. The Evolution and Future of the Human Animal. Huchinson, KS: Radius, 1991.

gets better with age. Loneliness is associated with early mortality. It was found that loneliness is worse for our health than smoking fifteen cigarettes a day. Social isolation and loneliness are associated with reduced levels of the neurotransmitter glutamate.[124]

6.8 The role of innovations and science in the human brain and social development

According to David Christian, our collective learning is a recent driver for what we are now. Through the magic of collective learning, humans learned how to increase their knowledge and skills. Cultural changes happen much faster than genetic changes. Farming was a mega-innovation. Humans can deliver a maximum of seventy-five watts of energy, while a horse or an ox can deliver up to ten times as much. These symbiotic relations can change the genetic makeup of both farmers and farming animals.

Christian also believed that the industrial revolution was another significant change for humans. New information flows stimulated European science and knowledge, fueling the industrial revolution. A major leap forward took place in the 1450s when Gutenberg invented and introduced printing. The Scottish inventor James Watts in 1765 progressed even further the industrial revolution. He added a second cylinder to the Newcomen engine invented by Thomas Newcomen in 1712 that was used primarily to pump water from coal mines. The added second cylinder served as a condenser, significantly increasing the power of the Newcomen

---

[124] Daniel Levitin. Successful Aging. A Neuroscientist Explores the Power and Potential of Our Lives. London: Penguin Publishing Group, 2020.

engine. Another big step in our industrial revolution happened in the middle of the First World War in 1916. That year, two German chemists, Haber and Bosch, invented how to condense air nitrogen, transforming it into an artificial fertilizer. This invention entirely changed agriculture, allowing food production for more people. It was suggested that without this invention, the nitrogen produced from soil could sustain agricultural production for only 3.5 billion people. Our thanks go to all scientists, like Harbor and Bosch, who helped humanity with their knowledge and innovations.[125]

Science and technology, governed by our individual and collective brain power, leads us to observe and discover new galaxies, new words, and new dimensions. I do not know how you feel, dear reader, but I feel very proud to witness this new history. Observing the stars through a telescope or directly makes me feel very responsible for the small piece of earth paradise that we have here all around us. I just keep looking at the sky, appreciating more and more what we have and how to help preserve it.

[125] David Christian. Origin Story: A Big History of Everything, London: Penguin, 2018.

6.9 The interaction of social and cultural factors with genetic factors

According to Kevin Mitchel, we are born with a pre-wired brain, but not hardwired. Brain plasticity leads to refinement of the neural circuitry in response to individual experience through the years. Genetics sets only the starting point. Humans have an extremely protracted period of brain development. It is valid especially for circuits that mediate behavioral control, such as in the prefrontal cortex; they do not fully mature until a person's early twenties. Most other animals have tightly defined ecological niches. They live in certain areas adapted to certain specific climates. They have a limited and highly characteristic set of behaviors. They lack the flexibility to adapt to new environments. At some point in evolution, the increasing ability to think in abstract terms led to, and was reinforced by, the emergence of language. The consequences were overpowering. Then children did not have to relearn everything new from their own experience. The culture was born. Cultural evolution began to collaborate with biological evolution. Intelligence gives us a huge advantage. We made our niche—the cognitive niche. Since the period of neural plasticity is so protracted in humans, we have the cognitive and neural capacity to continue to learn from experience over a much longer span of time.[126]

Joseph Henrich wrote that people are automatic cultural learners. The interaction between culture and genes (culture-gene coevolution) drove our species down

[126] Kevin J. Mitchel. Innate. How the Wiring of Our Brains Shapes Who We Are. Princeton, NJ: Princeton University Press, 2018.

a novel evolutionary pathway unobserved elsewhere in nature. Culture-gene coevolution can be remarkably fast. Our larger brains help us solve problems creatively. We excel in social learning. People are inclined to copy others more successfully. Humans are dependent on learning to survive. The author emphasized that our big brains are "energy hogs" using 20 to 25 percent of the energy we take in each day, while other primates' brains use between 8 and 10 percent. Natural selection has likely been at work, shaping our bodies for serious distance running for over a million years. We have enlarged gluteus maximus muscles, along with substantial spine muscles. We have large nuchal ligaments to sustain the head while running like other running animals. Our thermal adaptation is also extraordinary. Measured over body surfaces, no other animal can sweat faster than we do. Moreover, our eccrine glands are "smart glands" because they contain nerves that may permit centralized control from the brain. In other animals, sweating is controlled locally. Humans, unlike many mammals, can sustain core temperatures above 111.2 degrees Fahrenheit. Cooperative hunting and meat sharing were crucial elements in human evolution, reaching back millions of years into our past. Our ability to live in large groups depends heavily on social and cultural norms.[127]

Dr. Eric Kandel wrote, "We are who we are because of what we learn and what we remember." He believes that cultural evolution as a non-biological mode of adaptation operates side-by-side with biological adaptation as the

---

[127] Joseph Henrich. The Secret of Our Success. How Culture Is Driving Human Evolution, Domesticating Our Species and Making us Smarter. Princeton, NJ: Princeton University Press. 2016.

means of transmitting knowledge. His general idea is that learning and memory are central to our very identity. They make us who we are. Genes not only determine our behavior but also respond to environmental stimuli like learning. These complex interactions have been selected for over a million years by organisms struggling for survival and reproductive success. Our brains evolved to be complex, problem-solving machines. Our miraculous collective brain will help to resolve its own mysteries.[128]

One of the explanations for the complexity of our brain seems simple to understand. The scientific finding discovered recently and poorly popularized established that the brain neurons in primates have been increasing in number but not in size. In all other species, the brain evolved by increasing the size of the neurons. The evolutionary increase of our neurons by numbers gave us the huge advantage to have at present eighty-six billion neurons in each of our brains and far more connections between them. Now just imagine that we humans have the most neurons superimposed on our sophisticated astroglia cell system. One astroglia cell can connect up to ten thousand neurons and modulate their actions. Add the complexity of the microglial and oligodendroglial cell systems and you will start understanding the enormous complexity of our brains. This truly is an ocean of information and knowledge. Neuroscience needs total dedication and an influx of curious young scientists with active collaboration among all.

[128] Eric Kandel: In Search of Memory. The Emergence of a New Science of Mind. London, NY: W.W. Norton, 2006.

Pierre Teillard de Chardin, in his scientific and prophetic book *The Phenomenon of Man*, described a new evolutionary process: cultural evolution. Evolution is both the solution and the problem. The author believes there are no master navigators for our journey; we must navigate on our own. We must become wise managers of evolutionary processes. We need to adapt well to environmental changes, not only to a given environment. Culture, religion, and science all add value to our collective future. There is no dividing line between biology, human, and culture. Cultural evolution began to operate alongside genetic evolution, and the two processes started interacting. The products of cultural evolution enabled the human populations to adapt to their environments much faster than genetic evolution.

Dr. Michio Kaku emphasized that the mRIM-941 genes, discovered in Edinburgh in November 2012, are found only in *Homo sapiens*. These genes emerged between one and six million years ago. This series of genes perform in complex ways—no simple gene can do this. I was surprised when reading the book of physicist Michio Kaku, cited below. Despite being a physicist, he is deeply involved in the biology of human brain evolution. In his books, Dr. Kaku concluded that physics, biology, and cultures intertwine so intimately that all of us scientists need to collaborate. The human brain evolves at different angles and requires various specialists to resolve its deepest secrets.[129]

Dr. Nigel Barber confirmed that evolution is not restricted to genetic adaptation but works via learning

---

[129] Michio Kaku. The Future of the Mind. The Scientific Quest to Understand, Enhance, and Empower the Mind. NY: Doubleday, 2014.

and other mechanisms. Environmental stressors affect gene expressions. The author reports that children raised in stressful homes are significantly shorter in stature. Childhood abuse was associated with abnormal methylation in the adult brain, according to the analysis of suicide victims, especially of the glucocorticoid promoter. This can lead to decreased glucocorticoid receptor expression. Rat pups exposed to abusive maternal care have increased methylation of the BDNF gene in the frontal cortex. In humans, this methylation pattern is associated with major psychoses, including schizophrenia and bipolar disorder.[130]

Dr. Adam Rutherford explained the biological and cultural evolution interaction. We cannot focus only on "gene-centric" views of evolution. Biology drives culture and vice versa. We have a cultural transmission of skills. We accumulate knowledge and pass it on. Many animals learn. Only humans teach. Every journey of every human is built on thousands of years of accumulated knowledge.[131]

According to Nathan Lentis, the human brain is by far the most powerful cognitive machine on the planet. The advancement of the human brain beyond that of our closest relatives over the past six to eight million years has been truly exponential. One of the biggest mysteries is how humans became so much more intelligent than our closest relatives in such a short time. Thus, skills and social interactions were linked, and both evolved together,

---

[130] Nigel Barber. Evolution in the Here and Now. How Adaptation and Social Learning Explain Humanity. Amherst, NY: Prometheus Books, 2020.

[131] Adam Rutherford. The Book of humans. A Brief History of Culture, Sex, War and the Evolution of Us. London: Orion Publishing Group, 2019.

pushing the brain toward superior abilities. The rapid acceleration of brain growth in our ancestors was likely due to a switch to more competitive strategies for survival. How can such unspeakable monstrosity and genuine affection coexist in the same species? Our ancestors could switch between cooperation and competition whenever the conditions suited them. We evolved to be highly social, collaborative, and altruistic, but at the same time ruthless, calculating, and heartless. Evolution has made us extremely cooperative but entirely selfish too, just humans.[132]

In his book, Augustin Fuentes introduced the notion of "extended evolutionary synthesis," which includes genetic, epigenetic, behavioral, and symbolic inheritance (exchanging ideas). We became super cooperators. Hominins began to shape their world with collaboration and communication. They took this to the next level, trying to work together to get it done. New challenges grew our brains even more. By hunting animals, we better secured our food supplies. Only 5 percent of the chimps' diet is provided by hunting. Humans investigate and experiment beyond necessary for normal function and survival. Then we teach each other about it. Humans want to know why and how. We are particularly creative creatures. Creativity is both an individual and a group activity.[133]

Gaya Vince called the evolution of our genes, environment, and cultures our human evolutionary

---

[132] Nathan H. Lents: Human Errors: A panorama of our Glitches, From Pointless Bones to Break Genes. Boston: Houghton Mifflin Harcourt, 2018.

[133] Augustin Fuentes. The Creative Spark. How Imagination Made Human Exceptional. London, NY: Penguin Random House, 2017.

triad. The culture represents the learned information expressed in our tools, technologies, and behavior. Human cumulative culture over generations generates more efficient solutions to life's challenges. Cultural evolution shares much with biological evolution which operates mainly on the level of the individual, whereas for cultural evolution, group selection is more important than individual selection. The author wrote that our babies are born with a brain 28 percent of their adult size, while a newborn chimp's brain is 40 percent of the adult size. Humans need assistance in birth and especially after birth. Skull-delayed closure in humans allows two additional years of brain growth. Humans are dependent on social groups for survival. Cooperating, rather than competing, became highly important for our survival. Cultural evolution went hand in hand with technological development. Our cultural evolution is part of our biological evolution and vice versa.

Gaya Vance emphasized that our brains evolved with the reflexive use of narrative as part of our cognition. Stories shaped our minds, our societies, and our interactions with the environment. One study indicates that information shared by telling stories is twenty-two times more memorable. This is believed to be because multiple parts of our brain are activated as though we were living the story and experiencing it firsthand. Storytelling attaches emotions to the event and therefore makes stories memorable. Equally important is that our memory allows us to time-travel forward and imagine the future. To do this, our brain production system relies on a sophisticated type of memory that may be unique in humans: the episodic memory. Unlike most types of timeless memories, such as the ability to learn new skills or remember facts, episodic memory allows us to travel forward or back in time to visit any chosen event. Episodic memory, like language, relies on the cognitive connections between different brain regions. Apes do not have this ability, which evolved in our ancestors at least 1.6 million years ago. We know this because paleoanthropologists have discovered stone tools from that era that had been carried miles from their place of manufacture. Other primates do not plan when they acquire a surplus of food; they discard what they do not want momentarily. Food-storing animals like squirrels rely on behavioral instinct rather than conscious decision-making.

According to Dr. Vance, knowledge is the substance and the measure of cultural evolution. Cultures evolve too. This process is analogous to the mutation that occurs in genetic evolution. Cultural evolution progresses in leaps rather than by increments. Around five thousand years ago, humans invented the most brilliant and flexible information-storage tool: the written word, the key to cumulative cultural evolution. After all, the brain evolved as a prediction system to enhance its owner's survival. We are a unique coevolution of genes, environment, and culture. The role of language in humans is crucial too. Without language, we have no inner monologue, no system to arrange or formulate our thoughts. Music and language are both engendered in the same brain regions. Some linguists believe human speech began with a musical protolanguage. It was estimated that the global human population was around five million five thousand years ago. The population grew to be around 360 million by the time of the flourishing of the Silk Road around AD 800. Silk became a monetary exchange. Ancient roads with a total length of four thousand miles or so were built until the plague reached the Black Sea and two-thirds of the European population perished.[134]

In his wonderful book *Genghis Khan*, Jack Weatherford blamed the emergence of the plague in Europe for the establishment of regular long-distance communications in the Mongol Empire. The free movement of people and merchandise between Asia and Europe allowed the carriers of the disease, fleas from gray Asian rats, to invade, first the Italian merchant cities and later all of Europe mainly through ships arriving in the harbors.

> The authorities of the Italian cities realized the danger infection was connected to ships coming from Asia. They obliged ships to stay forty days at sea before entering the port. As all people speaking Latin-related languages know, *quarante* means "forty," and this is where the name "quarantine" comes from. Now we use the name "quarantine" for any span of time of isolation for infectious disease, such as seven, fourteen, or twenty-one days. History teaches; we learn.[135]

[134] Gaya Vince. Transcendence. How Humans Evolved Through Fire, Language, Beauty and Time, NY: Basic Books, 2020.

[135] For the ancient Babylonians, 7 is a sacred number. Their mathematics used base 7 instead of base 5. Thebes had seven gates. The year was and still is divided into weeks of seven days. Seven is sacred because it is made up of 3+4. *Three* is the number for the heavens and is doubly sacred, and *four* is symbolic of the earth (with, in classical antiquity, its four directions, four seasons, four winds, and four main rivers).

Fuller-Torrey explained that the hominin ancestral brain was estimated to be four hundred cubic centimeters some six million years ago. By the time of the Australopithecus, the brain volume was estimated to be 475 cubic centimeters. This appraisal was done by measuring the skull of Lucy, who was estimated to have lived 3.3 million years ago. Her skull and 60 percent of her skeleton were found in 1974 in Ethiopia, East Africa. The brain of *Homo habilis* 2.3 million years ago was estimated to be around 630 cubic centimeters. About 1.2 million years ago, hominids produced stone tools to open long bones and probably sucked the raw bone marrow. They could kill larger animals later and eat meat. At that time, the hominins were simply scavengers. During the era of *Homo habilis*, their brain grew to its present size. Humans began to exhibit more changes in the frontal and parietal lobes and started a new level of organization. Frontal lobes in humans contain four times more neurons than similar areas in chimps. Language development progressed, matching the growth of the frontal and parietal lobes. The language centers in apes are primitive and localized in the limbic system and the brainstem. Human language centers are in the frontal and parietal cortex. Interestingly, we also have a primitive one in the limbic system, but we use it only when we swear, cry, or laugh. Our language was an accelerant of human evolution rather than its cause.[136]

William von Hippel concluded that complex social relationships demand large brains. The social benefits of a big brain are enormous; it grew so large to solve

[136] Fuller-Torrey, Evolving Brains. Emerging Gods. Early Humans and the Origins of Religion, NY: Columbia University Press, 2017.

social rather than physical problems. *Homo sapiens* rose to worldwide dominance because of our hyper-sociability. The environment can seriously influence the way DNA is expressed and how this expression is passed on to future generations. Evolution appears to be a multifactorial combination of genes and environment.[137]

6.10 We can be the best helpers and the worst enemies

Brain Stetka wrote that our brains grew and were rewired to accommodate more collective living. According to the size of their brain, humans could maintain approximately 150 meaningful social ties. It is our large, adaptable, reorganized brain that led us to the world conquest. The group gives us protection, watchmen, and collaborators. Our brain is primed for the development of increased social behavior. Our amygdala fires when we freeze from fear and when we are anxious. The amygdala helps fix sensory memories into our brain for life. Emotions like fear and sex drive are the most essential and primitive mental states. Our fear promotes survival and our sex drive procreation, the two pillars of natural selection. The development of language was a major player in emotional evolution. We underwent a process of self-domestication through evolutionary adaptation that in our better moments allowed us to control the aggression. Our aggression can be associated with interhuman conflict, along with our wanton

[137] William von Hippel. The Social Leap. The New Evolutionary Science of Who We Are, Where We Came From, and What Makes Us Happy. NY: Harper Wave, 2018.

slaughter of animal species. In Europe and Asia, more than 50 percent of species are extinct and, in Australia, more than 70 percent. Our brains are so adapted to using the planet that we have created in doing so "a mess with rising tides and rising temperatures and without thousands of once thriving species."[138]

> We can be good, we can be bad, but good is still dominant in humans, at least for now. Thank you, human big brain!

The role of the X and Y chromosomes in our behavior: John Medina wrote that sex is set in concrete in DNA. Gender is not. In humans, the Y-chromosome decreased in genes' number at a rate of five genes every million years. So now we have only one hundred genes localized on our Y-chromosome that was called by scientists "suicide of the Y-chromosome in slow motion." By contrast, our X-chromosome contains more than fifteen hundred genes. The unusually high gene numbers in the X-chromosome were found to create proteins in brain manufacturing. The X-chromosome is called the cognitive hot spot because mutations in these genes often cause mental retardation and other brain problems.[139]

---

[138] Bret Stetka. A History of the Human Brain. From the Sea Sponge to CRISPR. How Our Brain Evolved. Portland, OR: Timber Press, 2021.

[139] John Medina. Brain Rules. Twelve Principles for Surviving and Thriving at Work, Home and School. Edmonds, WA: Pear Press. 2008.

Here is a recent story illustrating how observation in daily life inspires a desire to help. My wife and I were in the live audience at one of Dr. Phil's shows in the beginning of November 2019. I would like to share with you my clinical observation involving a young man and his girlfriend presented by Dr. Phil and his staff at this show.

Dr. Phil presented a twenty-three-year-old young woman who apparently had stopped her psychological development at age nineteen probably because of a major psychological trauma at that age. In front of the cameras, this young woman behaved like she was a nineteen-year-old. She had fallen in love with a little older unemployed man of a risky and unpredictable behavior. Among other hazardous behaviors, the couple stole a car and had an accident, causing significant human and other damage. The boyfriend was not at the show as he was already in jail. He was expected to be convicted to a maximum twenty-three-year prison sentence. The young woman was not out of trouble either. Dr. Phil and the lady talked to the jailed boyfriend on the phone. She continuously assured her boyfriend that she loved him very much and would wait for him until he served his time in prison. During this telecommunication, the producers showed only the picture of the boyfriend's. This picture stuck in my memory. I had seen several similar faces in my career as a child and later as an adult neurologist. This young man had big ears, approximately half the length of his face. He spoke smoothly, with no speech defect. A few seconds later, I could instantly connect his risky behavior and incapability to keep a job and his big ears to one syndrome related to a mutation in his X-chromosome I suspected the possibility that he may have fragile X-syndrome.

This syndrome is expressed clinically with big ears and different levels of mental handicap. The genetic mutation in this syndrome was found on an X-chromosome and can be proven with a relatively simple blood test. The psychiatric consultant for the show, Dr. Sophy, was involved in the case too.

My thought was that if they proved fragile-X syndrome in the jailed man, his risky behavior could be explained through that disease. In this case, he would benefit by being treated as a patient, not a criminal. I still believe that this jailed young man deserves medical attention before spending the next twenty years in prison. I believe we need to speak out, apply our experience, and try to help people wherever we are.

6.11 "We never stop learning because life never stops teaching." —Kirill Korshikov

As John Medina wrote, "I have never been able to turn off this fire hose of curiosity. We are natural explorers and lifelong learners. We need to explore and know. We can remain lifelong learners and problem-solvers. We lose thirty thousand neurons per day, but our brain continues to create new neurons, especially in the hippocampus, that show the same plasticity as those in newborns. Some parts of our brains stay as malleable as a baby's so we can create neurons and learn new things throughout our lives."[140]

I have also heard another saying appropriate for Medina's and my endless curiosity:

[140] Loc. cit. (see pp. 74 and 208)

Question: "What is the cure for boredom?"
Answer: "Curiosity."

Question: "What is the cure for curiosity?"
Answer: "There is no cure."

My curiosity led me to participate in many symposiums and science forums in my home country, Bulgaria, the USA, and all over the world. Yes, thank you, Kirill Korshikov, for the saying "We never stop learning because the world never stops teaching."

Below are summaries of some of my personal experiences as a curious neurologist interacting and studying with world-renowned specialists over the years:

1. I had the honor of being an interpreter for Dr. Andreas Rett when the Bulgarian Psychiatric Association invited him to see Bulgarian patients in 1988. Subsequently, in 1990, Dr. Rett invited me to be his lunch guest in his Vienna home. I will never forget the lunch he cooked for us that Saturday. His wife was away from Vienna at the time. Dr. Rett showed me his waiting room where, for the first time in 1964, he saw side by side two little girls rubbing their hands, as if washing them, a gesture typical of mental retardation. The two young girls were not verbal and had microcephaly (small heads). Their mothers tended them in their laps. Rett syndrome association in the USA is very active and helpful for all the Rett syndrome patients and their families who want to know and do more. As we know, we presently have a precision genetic test for this relatively rare but serious disorder.

2. Professor Shunsuke Ohtahara (1930-2013) described West syndrome in early infancy, a syndrome with frequent seizures and mental retardation. He is together with me on this picture in front of my poster at the International Muscular Dystrophy Symposium in Tokyo in 1991. Professor Ohtahara liked my CT imaging of the muscles of patients with Duchenne muscular dystrophy and especially the connection I made, detailed on the poster. The connection was that the muscles with the highest number of fibers innervated by one motor neuron degenerate first. For example, the gastrocnemius muscle (with thirteen hundred muscle fibers)—as compared to the anterior tibial muscle (with only three hundred muscle fibers)—degenerates much later. These data are still not published because the British journal to which I sent the article in 1991 requested many more radiological details.

As a practicing neurologist, I had no time to satisfy all the radiological requirements, so this paper could not be published. The next year, I came to the USA to start my medical career, and I had to abandon the pursuit of this publication. I hope some of the young investigators will continue my thirty-year-old work and idea.

3. In 1990, I attended the International Pediatric Neurosurgery Symposium in Japan. The organizers of that symposium had been interested in my clinical observation that idiopathic cranial hypertension, or the so-called pseudotumor cerebri and the Guillain-Barre syndrome, may represent two entities of a continuum of neurological syndromes. After the symposium, Professor Messay Segawa invited us to his parents' house located in the middle of the garden of the newly built Segawa Pediatric Neurology Institute.

Professor Segawa served us traditional Japanese tea. Professor Segawa shared with us that he is a seventh-generation physician in his family. He is well known in the medical world mainly because of the first-time description of dopa-responsive dystonia in 1971, now known as Segawa syndrome. In 1993, Professor Segawa sent me his book with an autograph.

4. The picture below shows Professor Sigvald Refsum in front of the Neurology Clinic, Sofia, Bulgaria, during the Fourth Bulgarian Congress of Neurology in 1988. Professor Refsum is well known mainly because of his description in 1946 of patients with polyneuropathy, ataxia, and retinitis pigmentosa, associated with high-level blood long-chain fatty acids, now known as Refsum syndrome. Professor Refsum honored our Bulgarian neurology clinic, especially his friends and my honorable neurology teachers, professors Sasho Bojinov and Ivan Gueorgiev. They created wonderful neuropathological and neurological traditions. These traditions are maintained at present by many of their students and my teachers and colleagues: professors Nachev, Shotekov, Nikoevski, Jordanov, Uzunov, Kililimov, Yanko T. Yankov, Kuchukov, Belopitova, Dora, and Bojinova. The younger generations are presented by professors Veneta Bojinova, Turnev, Avramov, and many others who continue their traditions and educate presently many new neurology professionals.

5. Dr. Colin Sullivan, from Australia, invented the CPAP (continuous positive airway pressure) in 1980. Before his discovery, the patients with severe sleep apnea were treated with a tracheostomy. The picture below was taken in front of our poster (with Dr. Ingmundson) during the first Sleep World Congress in Quebec, Canada, in 2008. Dr. Sullivan (in the middle) explained to us how he had the idea of using "blowing in through the device" for the first time. He had a patient with very severe sleep apnea who refused categorically a tracheostomy. To satisfy his patient's wishes and to save his life, Dr. Sullivan rigged a vacuum cleaner blowing in reverse through a handmade mask. Thus, 1980 became the birthdate of CPAP. Dr. Sullivan shared with us that during the next five years, he treated around a hundred patients with sleep apnea with his primary setup until a medical company bought his patent in 1985.

6. The Hospital Saint-Vincent-de-Paul functioning in this place in Paris from 1901 to 2011. I was lucky again to be specializing there in 1991–1992 in the neuropediatric department. All our thanks go to Professor Christophoroff, Dr. Rostaing, and the French government for providing

three-months scholarships for three Bulgarian physicians chosen by the Bulgarian medical authorities because of their French language skills. Professor Christophoroff and his wife, the Artiste Anne Christophoroff-de Colbert, opened their home and hearts for us. Thank you! My thanks also go to Professor Ponsot, Professor Du Lac, the neuropathology laboratory of the hospital, and all the staff of the hospital for sharing their advance knowledge that I brought to Bulgaria. My home country, Bulgaria, at that post-socialistic time needed all the help we could get.

7. The Italian Muscular Dystrophy Association under the guidance of Professor Neri invited several East European families with children suffering from muscular dystrophies. The meeting of the association was in the city of Salerno. The meeting organizers provided a bus for the families and their children and some meeting participants to visit Naples and see the antique city and Mount Vesuvius. I will never forget the patients' interest and the many questions they had.

Thank you also to the pioneer of all, the Muscular Dystrophy Association of USA, created in 1950 after the initiative of a New York business gentleman, Paul Cohen, affected himself with muscular dystrophy. Everybody will certainly remember the Labor Day Muscular Dystrophy Telethon, started in 1956, with the active participation of the actor Jerry Lewis. The telethon and many other donations helped the muscle diseases research tremendously. Thank you to all who are donating to medical research and certainly all researchers. We all have a lot more to do.

CHAPTER VII

How Our Complex Human Brain Can Benefit Our Future

The Universe is not all about us.
—attributed to Giordano Bruno

Learn how to "conquer
Nature by obeying her."
—Francis Bacon (1561–1626)

If everyone helps everyone, we
all can change the World.
—Leonardo da Vinci

We forgot we belong to one another.
—Mother Teresa

If there is no Vision,
there is no Hope.
—Dr. George Washington
Carver, agriculture professor

Earth provides enough to
satisfy every man's need, but
not every man's greed.
—Mahatma Gandhi

The biggest threat to the oceans is MAN.
—Fabian Cousteau, grandson of Jacques Cousteau.

Mark Berines wrote that our large brains and superior cognitive abilities permit us not only to survive but also to thrive. "We achieved top predator status. Violence has subsided but not inequality. This has led to erosion of human life support systems and creation of conflicts. We must return to our cooperative roots. We are not unique or exceptional even though our big brains fueled by cooked meat, group hunting, tool development, language and coevolution with other plants and animals have created this elaborate illusion. Will global habitat degradation be enough to trigger human cooperation to solve our planetary problems? What must we do next?"[141] Yuval Harrari concluded that the main source of wealth is knowledge. Knowledge is the most important economic resource. Humans crave more knowledge.[142]

> One of the latest developments for the future of gene therapy is the emergent CRISPR therapy. Here is the amazing story of CRISPR development and therapy. CRISPR is abbreviation for clustered regularly interspaced short palindromic repeats.

[141] Mark Beriness. A Brief Natural History of Civilization. Why a Balance Between Cooperation and Competition is Vital to Humanity. New Haven, CT: Yale. University Press. 2020.

[142] Yuval Noah Harari. Homo Deus. A brief History of Tomorrow. NY: Harper Perennial. 2017.

According to Walter Isaacson's interesting book *The Code Breaker*, CRISPR evolved in bacteria because of their long-running, three-billion-year-long war against viruses. This super interesting phenomenon was discovered and named CRISPR by a Spanish researcher in Alicante, Francisco Mojica, and described in 2005. This researcher, driven by curiosity, exclaimed in 2003 when he discovered this phenomenon, "Oh my goodness. Bacteria have an immune system. They can remember the viruses that had attacked them in the past." The fact that CRISPR targets the DNA of the invading virus was shown for the first time by Marraffini and Sontheimer in 2008. They suspected that CRISPR can be a genome-editing tool and can fix the cause of a genetic problem. They filed a patent application in September 2008 for "the use of CRISPR as a DNA-editing tool." However, their application was rejected with the explanation of the patent bureau that "you cannot patent an idea." They supported their negative answer based on the patent law written in Venice in 1474. "You can patent any new and ingenious device." Walter Isaacson explained in his book that for the last twenty to thirty years, scientists observed an interesting phenomenon in bacteria. If a hostile virus invades the bacterium, it uses its natural enzymes to cut off its own infected material, including the virus's genetic material. The aha moments arose in the brains of several scientists listed below to create similar mechanisms in mammals and ultimately in humans. They named this new trend in science and genetics CRISPR. Below are the institutions and scientists involved in this discovery described in Isaacson's book:

- Alicante University, Spain, 1990s: The graduate student Francisco Mojica observed thirty palindromic pairs in bacteria. Palindromic means that these thirty nucleotide letters can be read the same way from front to back and back to front. He also found that the spacers between these palindromic segments are thirty-six nucleotide pairs long. These spacers match some viruses that the bacterium is resistant to. Francisco Mojica called them CRISPR and found the phenomenon to be the system used by bacteria to defend themselves from viruses. Lately, it has been established that this mechanism of defense is a form of bacterial memory.

- Vilnius Institute, Lithuania, 2010: Virginius Siksnys found it possible to transfer the CRISPR system from one bacterial strain to another and confer viral resistance.
- RNA Symposium USA, 2011: Jennifer Doudna from Berkeley University, California, and Emanuel Carpenter from Vienna University, Austria, met and started talking. They visualized a molecule scalpel called CAS9 that could cut DNA. Both researchers received the Nobel Prize for biology in 2021. [143]

The future of science is bright, but we need to confront the risks of our present world first, the world we had created. Nighttime light pollution now affects more than a third of the earth's land area. For the first

[143] Walter Isaacson. The Code Breaker. Jennifer Doudna. Gene Editing and the Future of the Human Race. NY: Simon & Schuster. 2021.

time in history, we now live in physical isolation from our environment. What can we do?

> We need to resolve our problems. We are a significant part of these problems. It is vitally important to understand that none of us is the pinnacle of the universe. We need to understand one way or another that we are only one small part of our Mother Earth and the universe. Let us all do our part. Everybody needs to do his/her part according to his/her knowledge, capacities, and skills. Our complex brain and our social and technical skills have upgraded us to the status of "masters of our planet." For how long depends entirely on us. Let us start undoing the harms we have done to our planet. Let's keep our planet alive and well for many future generations. Yes, we owe this to future generations.

Stop destroying our birthplace, the only home humanity will ever have.

—Edward O. Wilson

This smiley earth was drawn by my oldest daughter, Svetla, at age sixteen when flying from Bulgaria to the USA to visit us in San Antonio. Let us keep our earth smiling for the generations to come. This book is my effort to help the younger generations. Please join me. Let us keep "growing" and interconnecting our precious human brains. Our humanity could be our survival and our immortality.

END

INDEX

C

D

M

N

W

X

Y

Z

Printed in the United States
by Baker & Taylor Publisher Services